Richard Price

Observations on Reversionary Payments

Richard Price

Observations on Reversionary Payments

ISBN/EAN: 9783337715793

Printed in Europe, USA, Canada, Australia, Japan

Cover: Foto ©Suzi / pixelio.de

More available books at **www.hansebooks.com**

OBSERVATIONS

ON

REVERSIONARY PAYMENTS;

ON

SCHEMES for providing ANNUITIES for WIDOWS, and for Persons in OLD AGE;

ON

The METHOD of Calculating the VALUES of ASSURANCES on LIVES;

AND ON

THE NATIONAL DEBT.

To which are added,

FOUR ESSAYS

On different Subjects in the Doctrine of LIFE-ANNUITIES and POLITICAL ARITHMETICK.

ALSO,

An APPENDIX AND SUPPLEMENT,

Containing additional Observations, and a complete Set of TABLES; particularly, several new Tables of the Probabilities of Life in different Situations, and of the Values of Annuities on Lives.

The THIRD EDITION, much ENLARGED.

BY RICHARD PRICE, D.D. F.R.S.

LONDON:
Printed for T. CADELL, in the Strand,
M.DCC.LXXIII.

TO

THE RIGHT HONOURABLE

THE

EARL of SHELBURNE,

THIS WORK is,

With all GRATITUDE and RESPECT,

INSCRIBED,

BY

His LORDSHIP'S

Moſt obliged, and

Moſt obedient humble Servant,

RICHARD PRICE.

CONTENTS.

PREFACE to the First Edition Page ix
Preface to the Third Edition p. xvii

CHAP. I.

Questions relating to Schemes for granting Reversionary Annuities, and the Values of Assurances on Lives — p. 1

CHAP. II.

SECT. I. *Of the London Annuity, and Laudable Societies for the Benefit of Widows.*
p. 64

SECT. II. *Of the Association among the London Clergy and the Ministers in Scotland, for providing Annuities for their Widows.*
p. 84

SECT. III. *Of the best Schemes for providing Annuities for Widows.* — p. 95

SECT. IV. *Of Schemes for providing Life-Annuities, which are not to commence 'till particular Ages; and, particularly, of the Societies lately established in London for the Benefit of Old Age.* — p. 106

SECT. V. *Of the* Amicable Corporation *for a perpetual Assurance-Office: and the Society for* Equitable Assurances on Lives and Survivorships. — — p. 120

CHAP. III.

Of Public Credit, and the National Debt.
p. 133

ESSAY I.

Observations on the Expectations of Lives; the Increase of Mankind; the Number of Inhabitants in London; *and the Influence of great Towns on Health and Population. In a Letter to* Benjamin Franklin, *Esq;* L.L.D. *and* F.R.S. p. 167. *To which is added, a Postscript, containing Observations on* Edinburgh, Paris, *and* Berlin.
p. 213

ESSAY II.

On Mr. De Moivre's *Rules for calculating the Values of* joint *Lives.* — p. 227

ESSAY III.

On the Method of calculating the Values of Reversions depending on Survivorships. p. 233

ESSAY IV.

On the proper Method of constructing Tables for determining the Rate of human Mortality,

CONTENTS. vii

lity, *the Number of Inhabitants, and the Values of Lives in any Town or District, from Bills of Mortality in which are given the Numbers dying annually at all Ages.*
p. 240

APPENDIX.

Containing Algebraical Demonstrations; Tables; and Rules for computing the Increase of Money bearing compound Interest. p. 283

SUPPLEMENT.

Containing additional Observations and Tables.
p. 357

information on this subject, I was led to undertake this work; imagining, that it might be soon finished, and that all I could say might be brought into a very narrow compass. But in this I have been much mistaken. A design, which I at first thought would give little trouble, has carried me far into a very wide field of enquiry; and engaged me in many calculations that have taken up much time and labour. I shall, however, be sufficiently rewarded for my labour, should it prove the means of preventing any part of that distress, which is likely to be hereafter produced by the societies now subsisting for the benefit of widows.——I have proved the inadequateness of their plans, by undeniable facts and mathematical demonstration.—I have, further, given an account of some of the best plans, that are consistent with a sufficient probability of permanency and success.——Should, therefore, any of these societies determine to reform themselves; or should any institutions of the same kind be hereafter established, they will here find direction and assistance (*a*).

In

(*a*) I have lately learnt, that Mr. *Cadell*, the publisher of this work, and also Mr. *Becket*, Bookseller in the *Strand*,

FIRST EDITION. xi

In Queſtion VI. Chap. I. a general method is deſcribed of finding the values, in
ſingle

Strand, are commiſſioned to deliver in *London*, printed accounts of the ſcheme of a ſociety, eſtabliſhed five years ago at *Amſterdam*, for granting annuities on ſurvivorſhip.—I cannot ſatisfy my own mind without introducing here, though an improper place, the following remarks on this ſcheme.

From the ſolution of Queſtions I. and IV. in the Firſt Chapter of the following Work, it may be gathered, that, (reckoning intereſt at $3\frac{1}{2}$ *per cent*. and the probabilities of life as they are in Tables III. IV. and V. in the *Appendix*) the value of an annuity of 1 *l.* for life, to be enjoyed by a perſon aged 20, provided he ſurvives another perſon aged 60, is 8 *l.* 16 *s.* 6 *d.* in one preſent payment; and 18 *s.* 6 *d.* in annual payments, during the two joint lives: the firſt payment to be made immediately. A *ſingle payment*, therefore, of 130 *florins*, entitles to an annuity of 15 *florins*; and an *annual* payment of 110 *florins*, to an annuity of 119 *florins*; and both together, to an annuity of 134 *florins*. If the annual payments are to be made, not during the *joint* lives, but during the whole continuance of the oldeſt *ſingle* life, they will, together with the ſingle payment, entitle to an annuity of 144 *florins*. But this ſociety promiſes, for theſe payments, an annuity of 100 *florins*, if the oldeſt life fails in the firſt year after admiſſion; 200 *florins*, if it fails in the 2d year; 300 *florins*, if it fails in the third; 400 *florins*, if it fails in the 4th; and 500 *florins*, if it fails in the fifth year, or at any time afterwards. It is, therefore, evident that the ſcheme of this ſociety is, in this inſtance, groſsly defective. There are other inſtances in which it is even more defective; and the *whole* of it, like the ſchemes of moſt of the *London* ſocieties, appears to have been contrived by perſons who had no principles to go upon. And yet it has been much encouraged. Many have entered themſelves into it from different parts of *Europe*; and the printed plan acquaints us, that it is now in poſſeſſion of an annual income

single and *annual* payments, of all life-annuities which are to begin after a given term of years; and, in the 4th Section of the 2d Chapter, the plans of the societies for granting such annuities are particularly considered, and proved to be extremely deficient.———
Indeed, the general disposition which has lately shewn itself to encourage these societies, is a matter of the most serious concern; and ought, I think, to be taken under the notice of the Legislature. The leading persons among the *present* members, will be the *first* annuitants; and they are sure of being gainers: and the more insufficient the scheme is, on which a society is formed, the greater will be the gains of the first annuitants. The same principle, therefore, that has produced and kept up other *bubbles*, has a ten-

come of 200,000 *florins*. What disappointment then must it in time produce?——It is provided by its rules, that the terms of admission shall become less and less advantageous, the longer it has subsisted; just as if the value of the annuities it promises depended, not on the probabilities of life, and the improvement to be made of money, but on the age of the society.——I have taken notice of a similar absurdity in the rules of our own societies. But it is easy to see what is meant by it.

Mr. *Cadell* can procure from his correspondents in *Holland*, any information for those who may want to know more of this society. But indeed I should be sorry to find it much enquired after in LONDON.

dency to preserve and promote these; and, for this reason, it is to be feared, that, in the present case, no arguments will be attended with any effect. The consideration, that " the gain made by some in these societies, " will be so much plunder taken from " others," ought immediately to engage all to withdraw from them, who have any regard to justice and humanity; but experience proves, that this argument, when opposed to private interest, is apt to be too feeble in its influence.

It cannot be said with precision, how long these societies may continue their payments to annuitants, after beginning them. A continued increase, and a great proportion of young members, may support them for a longer time than I can foresee. But the longer they are supported by such means, the more mischief they must occasion.—So, a tradesman, who sells cheaper than he buys, may be kept up many years by increasing business and credit; but he will be all the while *accumulating* distress; and the longer he goes on, the more extensive ruin he will produce at last.

In the latter end of the first Chapter, I have stated very particularly, the method of computing the values of *assurances* on lives and survivorships, in all cases where no more than two lives are concerned: and, in the 3d Essay, I have pointed out a considerable error, into which there is danger of falling in computing some of these values. The societies and offices for transacting business in this way, are very useful; and it is necessary that they should go upon the best principles, and possess all the information that can be given them.

But there is no part of this work in which the public is so much concerned, as the 3d Chapter. It will be there proved, that had the sums raised for public services since the REVOLUTION, been much greater than they have been, the increase of the public debts to their present state might have been prevented in the easiest manner, and at a trifling expence. A method, likewise, of reducing within due bounds these debts, heavy as they now are, will be proposed.—All competent judges will, I believe, see, that this method, being founded on the most perfect improvement that can be made of money, is the most

expe-

expeditious and effectual that the natures of things admit of. Nor, in my opinion, if the nation is not yet too near the *limit* of its resources, can there be any good reason against carrying it into execution.——— It is well known, to what prodigious sums, money, improved for some time at *compound interest*, will increase (*a*). A state, if there is no misapplication of money, must necessarily make this improvement of any savings, which can be applied to the payment of its debts. It need never, therefore, be under any difficulties; for, with the *smallest* savings, it may, in as little time as its interest can require, pay off the *largest* debts.

In the *first* Essay I have made many observations on the expectations of lives, the pernicious influence of great towns on health,

---

(*a*) A *penny*, so improved from our Saviour's birth, as to double itself every 14 years, or, which is nearly the same, put out to 5 *per cent.* compound interest at our Saviour's birth, would, by this time, have increased to more money than would be contained in 150 millions of globes, each equal to the earth in magnitude, and all solid gold. A *shilling*, put out to 6 *per cent.* compound interest, would, in the same time, have increased to a greater sum in gold than the whole *solar system* could hold, supposing it a sphere equal in diameter to the diameter of *Saturn*'s orbit. And the earth is to such a sphere, nearly as *half* a square foot, or a *quarto* page, to the whole surface of the earth.

and

and manners, and population; the increase of mankind; and other subjects in the doctrine of Annuities and Political Arithmetick——In the Last Essay I have stated carefully the proper method of forming Tables of the probabilities of human life, from given observations: And, in the *Appendix*, besides several new Tables, I have thought it necessary to give Mr. *Simpson*'s Tables of the values and expectations of LONDON lives; and all the other Tables which can be wanted in the perusal of this work.——I have also, in the *Appendix*, given the Demonstrations of the Answers to the *Questions* in Chap. I. These Demonstrations I have chosen to keep out of sight in the body of the work, in order to avoid discouraging such readers as may be unacquainted with mathematics.

Upon the whole, A great part of this work is, I believe, new; and I am in hopes also, that it will be found to contain some improvements in those branches of philosophical enquiry, which are the subjects of it.

PREFACE to the THIRD EDITION.

THAT favourable reception of this Work, which has occasioned the present Edition of it, so soon after two former editions, is such a proof that it has been of some use to the public, as amply rewards me for the attention and labour which I have bestowed upon it. In revising it on the present occasion, I have been anxious about improving it as far as possible. Several additional facts and observations have been inserted in different places, particularly in the first Essay and the Postscript to it.—That part of the second Section, Chap. II. which treats of the *Scotch* establishment, has been new composed, and carefully accommodated to the more accurate information concerning it, with which I have been favoured.—The 15th and 16th Tables in the *Appendix* are likewise additions, which I have

taken

taken this opportunity to make to this Treatise (*a*). The latter of these Tables gives the values of annuities on the *longest* of two lives, according to the mean probabilities of life, between *London* and the *Country*; and tho' these are values which every one may, without difficulty, calculate for himself, from the values given in Table VII. of *joint* lives, yet I have chosen to save those who use this work that trouble, and to lay before them in one view, the values of annuities on lives in all cases of *two* lives. The occasions for finding the values of annuities on *three* lives are much less frequent; and, therefore, I have thought no more necessary in this instance, than to recite at the end of the *Appendix* the rules by which they may, with ease and tolerable exactness, be determined.

The SUPPLEMENT is an addition which was made to the *second* edition.—The observations in it on the present state of our population I have enlarged and extended by a few notes; and, particularly, the *Postscript* beginning in page 379.—This is a very serious and important subject. If, indeed, there has been that diminution of our people which

---

(*a*) The three first Tables at the end of the *Supplement* have been also now first inserted in this work.

the

the evidence I have produced seems to prove, it must alarm every one who wishes well to his country, and it ought to engage the immediate and vigorous attention of government.——A well-known writer, Mr. ARTHUR YOUNG, and some other ingenious persons, differ from me on this point; and I wish I could be convinced by their arguments. But hitherto all my enquiries have served only to confirm me in my first conviction. Several great manufacturing towns have, I know, increased; but these are nothing to the whole kingdom; and even by their increase, our population may, on the whole, have lost more than it has gained.—— In truth; it would have been strange if our numbers had not been declining; for I can scarcely think of any great cause of depopulation, which has not for the last 80 years been operating among us. I think myself, however, obliged to Mr. *Young* for his remarks. The answer which I would give to the chief of them may be learnt from the notes in page 183, and 375 (*a*).

The last pages of the *Supplement* have been occasioned by accounts which I receiv-

(*a*) See likewise the second edition of the *Appeal to the Public on the Subject of the National Debt*, page 86, &c.

ed while this edition was in the prefs, and which came too late to be inferted in their proper places.

The prodigious traffic now carried on in Life-annuities, and the rage for forming and encouraging Annuity Schemes, which has for fome time been fpreading through the kingdom, has rendered the information which I have meant to convey in the following work particularly neceffary. And I have had the pleafure to obferve that it has been attended to. Several of the Annuity Societies in LONDON have been diffolved; and there is reafon to hope, that thofe which ftill remain will not be able much longer to fupport themfelves on their prefent plans, in oppofition to the evidence of demonftration, and the calls of juftice and humanity.——
Thefe *Bubbles*, however, are of little confequence, compared with that GRAND NATIONAL EVIL, which is the fubject of the fecond chapter of this treatife. This is an evil on which I could not imagine, that any fuch efforts as mine would make any great impreffion. Perhaps, indeed, the united efforts of all the independent part of the kingdom would now be too

weak

weak to save us from the distress with which it threatens us.

Much has been said for some time of a plan mentioned in PARLIAMENT, at the end of the last session, for paying off the NATIONAL DEBT. This raised some expectations; and, I will beg leave here to give a brief account of it.

After providing for all the current services, there remains this year a *saving* or *overplus* of 1,200,000*l*. With this sum, and a profit of 150,000 *l*. from a Lottery consisting of 60,000 tickets, (by a scheme similar to that described in the note, page 159, of the following work) a MILLION AND A HALF of the 3 *per cent*. annuities, purchased at 90, will be paid off (*a*).—When this was proposed to the House of Commons, it was at the same time declared, that it would be

(*a*) This scheme, applied to the purchase of the *Long Annuity* instead of the 3 *per cents*. would have gained considerably more for the public; and at the same time given equal profit to the stock-holders. The reason of this is, that the market price of the long annuity has for many years been constantly 5 or 6 *per cent*. below its true value, compared with the price of the 3 *per cents*.; so far, it seems, do the good people in the *Alley* look beyond 88 years, the present term for which this annuity is payable.

the COMMENCEMENT OF A PLAN FOR PAYING OFF THE NATIONAL DEBT; for, if no extraordinary fervices fhould call for any other application of the public furpluffes, the fame payment increafed by the intereft of former payments, is intended to be made every year while the peace lafts: And thus, reckoning compound intereft at 3 *per cent*. SEVENTEEN MILLIONS will be paid off during a peace of ten years.

On this plan I will take the liberty, with all the deference which becomes me to the ftation, abilities, and character of the propofer of it, to offer the following remarks.

1ft, It implies, that there is to be a *Lottery* every year during the whole continuance of peace.—Formerly, lotteries were expedients for procuring money on more advantageous terms, to which government had recourfe, when preffed by the neceffities of war. They are now, it feems, to be eftablifhed as *permanent* refources never to be given up or fufpended.—This muft fhock every perfon who is duly acquainted with the mifchief occafioned by lotteries, particularly among the lower claffes of people. The rage for gaming threatens the ruin of all

that

that is virtuous and manly among us. It is increasing fast, and wants not to be fostered by government.

2dly. The *surplus* of the present year is in part the effect of some *extraordinary* savings in the last year, which cannot be expected another year: And, I believe, that those who are best acquainted with this subject, must be sensible that there is no sufficient reason to expect, while the augmentation of the navy is continued, a constant *surplus* of so much as a MILLION *per ann.* I mean this on the supposition, that the produce of the *Sinking Fund* will continue what it is taken for this year, and what it has been the last three years, or 2,600,000 *l.* But this is certainly more than can be depended on. The difficulties of the *East India Company*; the stagnation of credit which has lately distressed the public, and many other causes, may possibly occasion *Deficiencies.* Should there, however, be even an *increase*, it will be owing, I am afraid, to a very bad cause: I mean, an increase of our importations proceeding from luxury, and turning the balance of trade against us; and, consequently, draining the kingdom of its *specie*, and leaving it more and more to the precarious and dangerous support of paper-money. But,

3dly, Let the *surplus* of the public revenue prove what it will, there is too much probability that, even during the continuance of peace, some emergencies or other will be often furnishing reasons or pretences for employing it in other ways than the payment of the public debts. This has been the case hitherto; and from the year 1730 to the present time, it has never happened, that we have gone on above three or four years together employing *surplusses* in discharging debts. Though in profound peace there have been calls for a different application of them; nor can I imagine what reason there is for believing, that our circumstances are so much changed for the better, that there will arise no such calls for ten years to come, should the peace last so long. But,

4thly, The most capital defect in this plan is, that its operation is to cease as soon as a war begins. That is; it is to cease at the very time when it would operate to most advantage, and when the greatest benefit might be derived from it. See this demonstrated in page 158 of this Treatise; and in p. 17 of *my Appeal to the Public on the Subject of the National Debt*.

Is it then any wonder, that such a plan has had no effect on public credit?—Does it mean any more than that the surplusses of the revenue shall be applied to the discharge of our debts, when there are no other uses for them?—And was there ever a time when this was not done? Is not this the very plan we have been pursuing these 40 years, and to which we owe our present incumbrances?—Certain it is, that nothing but a plan that shall go on operating uniformly in *war* as well as in *peace*, or the establishment of a permanent fund that shall never be diverted; that is, in other words, a return to the scheme adopted by the legislature in 1716; and which even now stands established by law, but which, through the unpardonable misconduct of men in power, has been defeated of its good effects: Nothing, I say, but this can do us any essential service; or, in our present circumstances, be much more than trifling with the difficulties and dangers of the public.— Establish such a fund—Consign it to a particular commission, acting under penalties, in such a manner as shall take it out of the hands of the *Treasury*, and form a check even on the *House of Commons* itself.—Supply

ply from time to time all deficiencies juft as if no fuch fund exifted; and, by thefe and other meafures, convince the kingdom that fomething effectual is meant, and that the public debts are indeed in the way to be extinguifhed.—LET THIS BE DONE; and we may foon fee a new ftate of things; public credit may revive; and the kingdom enjoy at leaft a chance for being preferved.—By the confidence which fuch a meafure would give in government fecurity; but, more efpecially, by the increafing fums which would be thrown annually into the public markets, and returned to the public creditors, the 3 *per cents.* would be foon raifed to *par*, and in fome time probably far above *par*. It is well known, what an effect *borrowing* every year has in finking the funds. Paying every year would certainly have an equal contrary effect. It would, to ufe the language of a very able writer on this fubject (*a*), caufe money to regorge in the hands of the lenders; and, with the help of prudent management, might be productive of confequences the moft advantageous.

In the interval of peace between the two laft wars, the 3 *per cents.* were at 105. Let

(*a*) Sir *James Steuart*, Bart. in his *Enquiry into the Principles of Political Oeconomy*.

us suppose that, in the circumstances I have mentioned, they would be raised to 110. Particular advantages might be derived from hence, which I will endeavour to point out distinctly, because, I think, they will shew in a striking light, how much might be done towards the extinction of our debts in a short course of years, were vigorous and STEADY measures entered into.

At the period I have supposed, instead of a reduction of *interest*, which would only retard the extinction of the public debts (*a*), the proper measure would be a reduction of the *capital*, attended with an advancement of interest, by such a measure as the (*b*) following.

The 3 *per cents* being at 110, and, consequently, an immediate loss of 10 *l.* arising to the proprietors from every 100 *l.* paid off, in order to prevent this loss, they would probably consent to a deduction from their capital of double this sum, provided what remained was made irredeemable for fifteen years, and the same interest continued—For,

(*a*) See this Treatise, page 139, &c.
(*b*) Since the above was written, I have found that a measure, in some respects similar to this, has been proposed by Sir James Steuart. *Principles of Political Oeconomy*, Vol. II. page 480.

1st, In this case they would submit for the present to no more than the imposition of a new name on their capital. That is, every proprietor of 100 *l.* stock being to receive 3 *l. per annum* for it, as he had always done, he would suffer only the inconvenience of hearing it called by the name of 80 *l.* stock (*a*).

(*a*) It deserves notice here, that such a measure as this has been actually employed to *increase* our debts.—In 1758, the lenders of 6,600,000 *l.* were entitled to a capital of 115 *l.* for every 100 *l.* subscribed, or of 7,990,000 *l.* in the stock of the 3 *per cent.* annuities: The consequence of which must be, that in discharging this debt, 15 *per cent.* or near a million, must be paid which was never received, and by which nothing has been gained.—This measure seems to have been adopted only to gain the *appearance* of borrowing at a low interest.—Were a person in private life to borrow 100 *l.* on the condition that it shall be reckoned 200 *l.* borrowed at 2½ *per cent.* he would, by subjecting himself to the necessity (if he ever discharged the debt,) of paying *double* the sum he had received, gain somewhat of the *air* of borrowing at 2½ *per cent.* though he really borrowed at 5 *per cent.* But would such a person be thought in his senses?—One cannot, indeed, without pain, consider how needlessly the capital of our debts has been in several instances increased.—I could shew, in particular, that about four millions of the consolidated 4 *per cents.* are an addition to the capital which has been made without the least reason for it, or the possibility of obtaining any advantages by it.—Thus do spendthrifts go on loading their estates with debts, careless what difficulties they throw on the discharge of the principal, leaving that to their successors, and satisfied with any expedients that will make things do their time.—When will our Statesmen learn to carry their views to futurity?

But,

But, 2dly, The difcharge of the capital being not to take place till after the expiration of 15 years, and then only to commence and to be the gradual work of feveral years, the benefit offered to the public creditors would, in reality, be near the true value of the reduction to which they confented.—For inftance—20 *l.* the payment of which is to be delayed fifteen years, and then to be made by fmall annual payments till completed, cannot be worth in prefent money much more than 10 *l.* and, therefore, it would be reafonable in the proprietors of a 100 *l.* ftock to give up 20 *l.* for it on fuch terms, in order to fave 10 *l.* in hand.

But it feems certain, that, in the circumftances I am fuppofing, the public creditors would be glad to give up a larger fum than was equivalent to the value of the prefent fum faved. For, the lofs being future and diftant, it would, in confequence of principles neceffary in human nature and often fatally prevalent, be much lefs regarded than in proportion to its true value.

But, farther; this lofs would be confidered in general as a lofs likely to fall on pofterity, or fome *future* purchafers of ftock, and

and not on any *present* creditors; and, consequently, the same disposition that has formed and promoted the bubbles which have done so much mischief in this kingdom, would, in this case, be made to operate to its advantage.

I have, therefore, certainly kept within bounds, when I have reckoned that a reduction of 20 *l. per cent.* in the capital of the 3 *per cents.* might be made, in the circumstances I have mentioned.——Let then such a reduction be supposed to be applied to *sixty millions* of the 3 *per cents.* This will leave much more than enough free for the operations of the fund; and by such management as that, which, in 1749, reduced 57 millions from an interest of 4 *per cent.* to an interest of 3 *per cent.* there is no reason to doubt but it might be accomplished in *one* year, or at most in *two* or *three* years; and the consequence would be, that a capital of sixty millions would be reduced to 48 millions; or, that twelve millions of debt would be cancelled without expence or difficulty.

But this is not the only advantage which would arise from such a measure.——

At

At the end of the term I have mentioned, 48 millions would be *redeemable* debts, bearing 3¼ *per cent.* intereſt. Theſe would ſell much above *par*; and a *ſecond* reduction, on condition of irredeemableneſs for a *ſhorter term*, might be applied to ſuch a part of them as it might not be neceſſary to leave free; and thus, by the ſame means with the foregoing, ſeveral millions more might be annihilated.—At the ſame time the fund, which had hitherto been employed in diſcharging redeemable 3 *per cents*, might be applied to the diſcharge of debts bearing 3¼ *per cent*. intereſt, and therefore would, as proved in page 138, be accelerated in its operation. And at the end of the *ſecond* term, it might be applied to debts bearing a ſtill higher intereſt, and therefore would be ſtill more accelerated.——This ſeems to go to the very limit of poſſibility on this ſubject.—Money in a fund, NEVER DIVERTED, is improved at compound intereſt; and, this being the very beſt improvement of money poſſible, there can be no method of diſcharging debts ſo expeditious. But by the ſcheme now explained, THE OPERATIONS OF COMPOUND INTEREST ITSELF WOULD BE AIDED. It

would

would be eafy to fhew, that, in 40 years, and without the aid of *Lotteries*, a HUNDRED MILLIONS of the 3 *per cents.* might in this way be difcharged, with a prefent annual furplus of (*a*) no more than 900,000 *l.* to be increafed in the year 1781 by 200,000 *l.* (*b*) which the public will gain by the reduction of the confolidated 4 *per cents.* to 3 *per cents.* And this, without all doubt, is near TWICE as much as can be done in the fame time with the fame *furplus*, by any other equitable means.—With a prefent annual *furplus* of a *million*, no more than *twenty-five* millions of the 3 *per cents.* would be converted into life-

(*a*) About twenty millions would be difcharged without any difburfement of money; and the remainder would be difcharged by the accumulation of the *fund*, applied, for the firft 25 years, to the payment of debts bearing 3 *per cent.* intereft, and afterwards to debts bearing higher interefts.

The management above propofed might be applied to the propofal in page 156, and would very much improve it.—That propofal requires a prefent furplus of a million and a half *per annum*; and could fuch a furplus be gained, our deliverance would be rendered much more probable and complete; but that being more than can be obtained without retrenchments and favings, which, however practicable, are not to be expected, I have been induced to enquire what might be done with fmaller furpluffes.

(*b*) In 1782 there will be another faving gained, from the reduction of *four millions and a half*, $3\frac{1}{2}$ *per cent.* annuities, 1758, to an intereft of 3 *per cent.*

annuities, suppofing the proprietors, one with another, to accept, in lieu of every 100 *l.* ftock, 7 *l. per ann.* for life. And the whole incumbrance on the public occafioned by fuch annuities, would not be entirely removed in lefs than *feventy*, or perhaps *eighty* years.

Were a furplus of a *million per ann.* employed in converting the 3 *per cents.* into *long annuities*, a hundred millions might indeed be difcharged, by locking it up for a term of years, and offering the proprietors 4 *l. per ann.* for that term, in lieu of every 100 *l.* ftock. But it would be neceffary to make the term much longer than *forty years*. He that will confider the low price of the *long annuity* now at market, may fatisfy himfelf, that no term fhorter than *fixty* or *feventy* years would be accepted; and the fame *furplus*, locked up for *feventy* years, would, in the way I have propofed, difcharge THREE HUNDRED MILLIONS.

I muft repeat here what cannot be too much inculcated, that a war would have no other effect on fuch a fcheme than to aid it. The operations of the fund would be quickened

ened in the manner explained in page 157, &c. And, suppofing no diverfions of it during the exigences of war, fuch a demonftration would be given to the public, that an *unalterable* plan was at laft eftablifhed, as could not fail to produce the happieft effects; and to enable government, when peace came, to carry into execution fuch meafures as I have propofed to the greateft advantage.

The lofs of the million *furplus*, in a time of war, is a lofs that muft be fubmitted to, whatever plan is adopted; nor would it, in that which I have propofed, be productive of any additional burdens or difficulties. —In war it would be neceffary to borrow feveral millions annually; and, at fuch a time, the neceffity of borrowing *one million extraordinary* could not make any great difference: And, as this would be done to convey a conviction with which the very power of borrowing was connected, and to preferve a *fund* on which the very being of the ftate depended, none but the beft confequences could arife from it. The public burdens would be even *lefs* increafed by a war, in
con-

consequence of having a million *per annum* during its continuance, thus withdrawn from the supplies. For, let us suppose *six millions* necessary to be borrowed every year to defray the expences of war, *five millions* only of which would have been wanted, had not the million *surplus* been locked up.——Suppose farther, that the scheme, by keeping up public credit, and throwing money every year into the hands of lenders, enables government to borrow at 1 *l. per cent.* less interest than would be otherwise required, or at 4 instead of 5 *per cent.*—In these circumstances, there would arise a present saving to the kingdom of 10,000 *l. per ann.*; for the interest of *six millions* at 4 *per cent.* is 10,000 *l.* less than the interest of *five millions* at 5 *per cent.* (*a*).

And

(*a*) There would, indeed, be an increase of *capital*; but this we have hitherto never regarded, when it has not been attended with an increase of *interest*. In the present case, however, it would not be necessary, that the increase of capital should make any addition to the public burdens. For,

1st, The scheme might soon be applied to the capital, and would cancel it faster than the capital of 3 *per cents.* on account of the higher interest it bore.

2dly, The price of it would, when peace came, rise far above *par*; and, therefore, it might easily be reduced

from

And such a saving, repeated every year of a war, would be an object of some importance to the kingdom.—Indeed, there may be no possibility of conceiving what important effects in this way, the establishment of such a scheme might produce. During its progress in discharging our debts, and before it could give any relief by the annihilation of taxes, it might *save* the kingdom, by preserving it from difficulties which would have sunk it. And every one must be sensible of this, who has considered what danger there is that a war, should it become unavoidable before our debts are put into any certain method of redemption, will either entirely overwhelm public credit, or so much weaken it, as to produce an impossibility of borrowing, except on very exorbitant interest, and, consequently, of finding taxes sufficiently productive to pay such interest.—The gene-

---

from six to five millions by the management I have explained.

3dly, There are even methods by which *six millions* might be borrowed at 4 *per cent.* and the capital fixed, without inconvenience or difficulty, *to five millions.*—— Those who do not chuse to give me credit for this, may, if they please, think it a mistake. The full explanation of it would lead to an account of the best method of contracting debts, for which I have here no room.

ral

ral apprehenſion now is, that the nation is overloaded; and that its debts will never be paid. This keeps the funds near 18 *per cent.* lower than they were in the laſt *peace*. In the next war ſuch apprehenſions will increaſe, and produce great danger. But ſhould it be then ſeen, that a plan for redeeming our debts, the moſt efficacious that poſſibility itſelf allowed, was going on; and, in conſequence of being guarded in ſome ſuch manner as I have hinted, *would* not, or *could* not *eaſily*, be revoked; in theſe circumſtances, all danger would be ſo far leſſened, that it might be practicable to find new taxes which would ſupport the expences of war during the operations of the ſcheme.—If any one believes the contrary, let him, in God's name, think what a condition we are in.—I hope our circumſtances are not ſo deſperate.——Many ſavings might certainly be made, without particular difficulty, in the collection and expenditure of the revenue.—A conſiderable annual income might be derived from taxes upon *horſes*, *dogs*, *livery-ſervants*, and *celibacy*; from an increaſe of the tax upon *coaches* and *plate*; and from a tax on all legacies and successions to estates. The last tax would be only obliging those who had enjoyed the protection of the state during life, to contribute towards its support at death. And all the other taxes would necessarily do good in whatever way they operated.

But I am got far beyond the limits I prescribed myself when I begun this Preface.— As the national debt is a subject unspeakably interesting (a) to this nation, I could not allow myself to omit any thing that appeared to me of consequence upon it; and the Reader of this Treatise will on this account, I hope, excuse me, if I have detained him here too long and too improperly.—In reviewing what I have written, I am indeed almost disposed to congratulate myself on having pointed out a method of discharging the public debts in a short period of years,

(a) Mr. GORDON tells us, that the great and good Mr. TRENCHARD had two things much at heart, namely, keeping *England* clear of foreign broils, and paying off the public debts. He thought that one of these depended on the other, and that *the fate and being of the State depended on the latter*. Mr. GORDON adds, that he believed no one who thought at all, could think Mr. TRENCHARD mistaken. Preface to Cato's Letters.

with a *surplus* now in our possession, and the INVIOLABLE appropriation of which will *never* be felt, except in effects the most salutary and beneficial.—But I fall back into diffidence. Much has been before said on this subject by writers of more consequence to no purpose; and we shall pursue the path we are in, till the edge of the precipice towards which we are advancing awakens us, and ruin becomes certain and unavoidable.—The distress occasioned by the shock lately given to the bubble of paper-credit, is, I am afraid, a prelude to unspeakably greater calamities, and a warning to prepare for them.

ERRATA.

ERRATA.

Page 41. *line* 6. *for* marriage in seven fails of leaving children that survive their parents, *read* one in seven of all who die widowers leave no children.

Page 79. *line* 14. *for* exceed considerably the number of marriages. *read* exceed considerably *half* the number of marriages.

Page 316. *column* 3. of the first Table, *line* 1 from the bottom, *for* .0199, *read* .0899.

CHAP. I.

Questions relating to Schemes for granting Reversionary *Annuities, and the Values of Assurances on Lives.*

QUESTION I.

"A Set of married men enter into a
" society for securing annuities to
" their widows. What sum of
" money, in a single present pay-
" ment, ought every member to contribute,
" in order to entitle his widow to an annuity of 30 *l. per ann.* for her life, estimating interest at 4 *per cent ?*"

ANSWER.

It is evident, that the value of such an expectation is different, according to the different ages of the purchasers, and the proportion of the age of the wife to that of the husband. Let us then suppose, that every person in such a society is of the same age with his wife, and that one with another all the members when they enter may be reckoned

oned 40 years of age, as many entering above this age as below it. It has been demonstrated by Mr. *De Moivre* and Mr. *Simpson*, that " the value of an annuity on the *joint con-* " *tinuance* of any two lives, subtracted from " the value of an annuity on the life in ex- " pectation," gives the true present value of an annuity on what may happen to remain of the latter of the two lives after the other.

In the present case, the value of an annuity to be enjoyed during the *joint continuance* of two lives, each (*a*) 40, (*b*) is 9.826, according

---

(*a*) See Table VII. Appendix.

(*b*) The values of *joint* lives and reversions, as deduced from the *Breslaw* observations, are not given in any part of this work from Mr. *De Moivre*'s rules in his treatise on annuities on lives. For these rules are approximations, which give results so far from the truth, as to be, not only useless, but dangerous. In the second essay in the Appendix, a particular account of this will be given, and also of the method in which these values have been calculated.

Mr. *De Moivre* has calculated the values of *single* lives, on the supposition of an *equal decrement of life* thro' all its stages till the age of 86, which he considered as the utmost probable extent of life. Thus; let there be 56 persons alive at 30 years of age. It is supposed that one will die every year till, in 56 years, they will be all dead. The same will happen to 46 at 40, in 46 years. To 36 at 50, in 36 years, and so on for all other ages. The number of years which a given life wants of 86, he calls the *complement* of that life. Fifty-six, therefore, is the *complement* of 30; 46 of 40, and 36 of 50.

This hypothesis eases very much the labour of calculating the values of lives; and it is so conformable to Dr. *Halley*'s table of observations, that there is little or no reason

ing to the probabilities of life in the Table of Observations formed by Dr. *Halley*, from the bills of mortality of *Breslaw* in *Silesia*. The value of a single life 40 years of age, as given by Mr. *De Moivre*, agreeably to the same Table, is 13.20 (*a*); and the former subtracted from the latter, leaves 3.37, or the true number of years purchase, which ought to be paid for any given annuity, to be enjoyed by a son for distinguishing between the values of lives as deduced from this Table, and the same values deduced from the hypothesis.

In order to avoid putting the reader to trouble, I have given this table at the end of this work. And I have also given two other tables which I have formed from the bills of mortality at *Northampton* and *Norwich*. These last tables answer more nearly to Mr. *De Moivre*'s hypothesis than even Dr. *Halley*'s table; and the difference between the values of *single* and *joint* lives by the *hypothesis*, and the same values computed strictly from the tables, is generally less in these tables than in Dr. *Halley*'s, as will be shewn in the last Essay. When, therefore, in the course of this work the values of *single* and *joint* lives are mentioned, as given agreeably to Dr. *Halley*'s table, it must be understood, that they are taken from Tables VI. and VII. in the Appendix, and given in strict agreement only to the *hypothesis*; and that for this reason, they are in reality still more conformable to the *Northampton* and *Norwich* tables.

The inhabitants of *London*, as is well known, not living so long as the rest of mankind, the values of *single* and *joint* lives there, are considerably less than those just mentioned. And, therefore, whenever I have had *London* lives in view, I have given particular notice of it; and taken their values from Mr. *Simpson*, who has calculated them with much accuracy from the *London* tables of observation. See Tables X. and XI.

(*a*) See Table VI. Appendix.

person 40 years of age, *provided* he survives another person of the same age, interest being reckoned at 4 *per cent. per annum.* The annuity, therefore, proposed in this Question being 30 *l.* the present value of it is 30 multiplied by 3.37, or 101 *l.* 2 *s.*

By calculating from Mr. *Simpson*'s Tables *(a)*, formed from the bills of mortality of *London,* this value comes out 102 *l.*

The difference in the value of the reversion will be inconsiderable, whether the common age is taken a few years more or less than 40. Thus married men of 30 ought not, according to Dr. *Halley*'s Table, to give two-fifths of a year's purchase more, for any given reversionary annuity for their wives, than married men of 50, provided they are of the same ages with their wives; and one quarter more, according to Mr. *Simpson*'s Table. If the wives are younger (as is generally the case) there will indeed be a considerable difference; for the value now determined would be 120 *l.* according to the *Breslaw* Observations, supposing the two lives to be 40 and 33, or that wives are one with another seven years younger than their husbands; and 118 *l.* 10 *s.* according to the *London* Observations.

(a) See Table X. and XI. Appendix.

Question II.

"Supposing such a society as that described in the preceding Question, to be limitted to a certain number of members, and constantly kept up to that number, by the admission of new members as old ones are lost, in consequence of their own deaths, and the deaths of their wives: What is the number of annuitants which, in some time after its establishment, will come to be constantly upon it?"

Answer.

Since every marriage produces either a widow or widower; and since all marriages taken together would produce as many widows as widowers, were every man and his wife of the same age, and the chance equal which shall die first; it is evident, that the number of widows that have ever existed in the world, would, in this case, be equal to *half* the number of marriages. And what would take place in the world, must also, on the same suppositions, take place in this society.——In other words; every *other* person in such a society leaving a widow, there must arise from it a number of widows equal to half its own number.——But this does not determine what number, all living at one and the same time, the society may expect will come to be constantly upon it. For if every widow lived no more than a year, the society would never have more annuitants upon it, than came on in a year. And on the contrary, if none ever died, the number of annuitants would go on increasing for ever.—'Tis, therefore, necessary, in order to answer the present enquiry, to determine how long the *duration* of *survivorship* between persons of equal ages will be, compared with the *duration* of *marriage*. And the truth is, that, supposing the probabilities of life to decrease uniformly *(a)*, the former is equal to the latter; and consequently, that the number of *survivors*, or (which is the same supposing no second marriages) of *widows* and *widowers* alive together, which will arise from any given set of such marriages constantly kept up, will be equal to the whole number of marriages; or *half* of them (the number of widows in particular) equal to *half*

(a) That is, supposing that out of any given number alive at any age, the same number will die every year 'till all are dead. See the preceding note. That on this hypothesis, the duration of survivorship is equal to the duration of marriage, when the ages are equal; or, in other words, that the *expectation* of two joint lives, the ages being equal, is the same with the *expectation* of survivorship, may be learnt from the 18th and 20th problems of Mr. *De Moivre*'s treatise on annuities; and a demonstration of it, together with a particular explanation of this subject, may be found at the beginning of the first Essay, to which I must beg the reader to turn, if he is at any loss about the full meaning of what is here said.

the

the number of marriages.—Now, it appears that the decrease in the probabilities of life, is in fact nearly uniform. According to the *Breslaw*, the *Northampton* and *Norwich* Tables of Observation, almost the same numbers die every year from 20 years of age to 77 *(a)*. After this, indeed, fewer die, and the rate of decrease in the probabilities of life is retarded. But this deviation from the hypothesis is inconsiderable; and its effect, in the present case, is to render the duration of survivorship *longer* than it would otherwise be. According to the *London* Table of Observations, the numbers dying every year begin to grow less at 50 years of age; and from hence to extreme old age, there is a constant retardation in the decrease of the probabilities of life *(b)*. Upon the whole, therefore, it appears in answer to the present Question, that " according to the *three* " *former Tables* of Observations, and suppo- " sing no widows to marry, the number " enquired after is *somewhat greater* than " half the number of the society; but, ac- " cording to the *London Table*, a *good deal* " *greater*."

It must be carefully remembered, that this has been determined on the supposition, that

(a) See Tables III. IV. and V. Appendix.

(b) The reason of this difference between the *London* and other Tables, will be given at the end of the fourth Essay.

husbands and their wives are of equal ages, and that in this case it becomes an equal chance which shall die first. In reality neither of these suppositions is just. Husbands in general are older than their wives; and, in equal ages, the mortality of males has been found to be greater than the mortality of females. For both these reasons, it is much more than an equal chance that the husband will die before his wife, or that the woman shall be the survivor of a marriage, and not the man. This will increase considerably the duration of survivorship on the part of the woman, and consequently the number enquired after in this Question. The marriage of widows will also diminish this number, and the operation of these causes will be different in different situations. But it is by no means to be expected (in the situation of the societies I have in view) that the diminution from the latter cause will be considerable enough, to overbalance the operation of all the other causes which have been mentioned, and reduce the number under consideration so low, as half the number of marriages *(a)*.

SCHOLIUM.

In *London* it appears, that there is a retardation of the decrease in the probabilities

---

(a) It will be observed hereafter, that this observation has been found to be true in fact.

of life, which renders the duration of survivorship between two lives of equal ages, considerably longer than their joint continuance. It seems worth observing, that this is the reason why, though the probabilities of life, and therefore the values of single and joint lives, are less in *London* than in other places, yet the values of reversions depending on survivorships, are in some cases greater there. It is proper to add, that this likewise is the reason why, in calculating the values of joint lives and reversions, the present value of an annuity payable yearly to the survivor of two equal lives, may come out equal to, or even greater than, the present value of a like annuity for the joint lives. As an annuity, during such survivorship, will probably not become payable for some years, and therefore the money given for it will have time to accumulate, it is manifest, that the value of it could never be equal to the value of an annuity on the joint lives, the payment of which begins immediately, were not the observation now made true.

Question III.

" Such a society as that described in the
" preceding Questions being supposed; in
" what time will the number of annuitants
" upon it come to a *maximum*?"

ANSWER.

In order to be more clear in answering this Question, I will first suppose the society to comprehend in it from its first establishment, *all* the married persons of *all* ages in any town or country, where the number of people continue constantly the same. In this case, the whole collective body of members will be, at their greatest age, at the time of the establishment of the society; and the number of members, together with the number of widows left every year, will, taking one year with another, admit of no increase or diminution. The number of widows in life together, derived from any given number coming on a society every year, will increase continually, 'till as many die off as are added every year; that is, 'till they come to die off as fast as possible. But they cannot die off as fast as possible, 'till the whole collective body of widows are at their greatest age; or, 'till there is among them the greatest number possible of the oldest widows; and, therefore, not 'till there has been time for an accession to the oldest widows, from the youngest part of the widows that come on annually.

Let us, for the sake of greater precision, divide the whole medium of widows that come on every year, into different classes according to their different ages, and suppose some to be left at 56 years of age, some at 46, some

some at 36, and some at 26. The widows, constantly in life together, derived from the first class, will come to their greatest age, and to a *maximum*, in 30 years, supposing with Mr. *De Moivre*, 86 to be the utmost extent of life. The same will happen to the second class in 40 years, and to the third in 50 years *(a)*. But the whole body, composed of these classes, will not come to a *maximum*, 'till the same happens to the fourth or youngest class; that is, not 'till the end of 60 years. After this, the affairs of the society will become *stationary*, and the number of annuitants upon it of all ages will keep always nearly the same.

Such is the answer to this Question, supposing a society to begin with its complete number of members, consisting of married persons of all ages, in the same proportions to one another, with the proportions in which they exist in the world.——If it begins with its complete number of members, but at the same time admits none above a particular age: If, for instance, it begins with 200 members all under 50, and afterwards limits itself to this number, and keeps it up by admitting every year, at all ages between 26 and 50, new members as old ones drop off;

(a) In the *Appendix*, note (A), a rule is given, by which the numbers alive at the end of any particular number of years may be very easily determined.

in this cafe, the period neceffary to bring on the *maximum* of annuitants will be juft doubled. For, in the firft place, the whole collective body of members will be 60 years in getting to their greateft age, as may eafily appear from what has been juft faid. The annual medium of widows, therefore, that will come on the fociety will increafe continually for 60 years; it being evident, that the older any fet of married men are, taken one with another, the fafter they will leave widows. And after this annual medium is increafed to a *maximum*, 60 years more will be neceffary to bring to a *maximum* the number in life together, derived from fuch a *fixed* annual medium conftantly coming on.——If fuch a fociety is any number of years in gaining its *maximum* of members, the time neceffary to bring on the *maximum* of annuitants will be ftill further prolonged, and will be equal to twice 60 years with that number of years added.——Moft of the focieties for granting annuities to widows are of this kind; and, therefore, fuppofing them to gain their complete number of members in ten years, and for ever afterwards to preferve it, the number of annuitants upon them will go on increafing for 130 years.——It is proper, however, to be remembered, that the increafe will be quicker at firft, and afterwards flower; and that, within 20 or 30 years of the end of of this term, it will be so slow as scarcely to be sensible, though still real.

All who will bestow due attention on this subject must see these decisions to be just; and a demonstration of them might be given, in a form more strictly mathematical, were it necessary.

Question IV.

"Suppose the members of such a society as that described in the preceding Questions, to chuse making *annual payments during the continuance of marriage*, in lieu of the sum which the reversionary annuity for their widows is worth in *present money*: What ought these *annual payments* to be, estimating interest at 4 *per cent*?"

Answer.

This will be easily determined, by finding what annual payments, during two joint lives of given ages, are equivalent to the value of the reversionary annuity in *present money*.— Suppose, as in Question I. the two joint lives to be each 40, and the reversionary annuity 30 *l. per annum*. An annual payment during the continuance of two such lives is worth, according to Dr. *Halley*'s Table of Observations, 9.82 (*a*) years purchase. The annual

(*a*) See Table VII.

payment then ought to be such as being multiplied by 9.82, will produce (*a*) *l*. 101.1, the present value of the annuity in one payment by Question I. Divide then *l*. 101.1 by 9.82, and the *quotient*, or *l*. 10.3 will be the answer.——This is very nearly the annual payment of all the members at an average, supposing equal numbers to offer themselves for admission of every age between 30 and 50. As much as some give less, others ought to give more, according to their excess of age. Thus, the annual payment of a married person, 30 years of age, ought to be *l*. 9.39; and of a person 50 years of age *l*. 11.33.——If the values of joint lives and of the reversionary annuity are taken agreeably to the *London* Table of Observations, these annual payments will be, for 30 years of age (*b*), *l*. 10.9,—for 40, *l*. 12.5,—for 50, *l*. 14.5.

If

(*a*) Particular notice should be taken of the method of notation here used, because it will be carried through the whole of this work.——The figures on the right hand of the full-point, signify the decimal parts of 1 *l*. Thus; *l*. 101.1, is 101 and the 10th of 1 *l*. or *l*. 101 and 2 *s*.——*l*. 9.39, is *l*. 9, and 39 hundredths of 1 *l*. or *l*. 9 : 7 *s*. : 10 *d*.——*l*. 11.33, is *l*. 11, and 33 hundredths of 1 *l*. or *l*. 11 : 6 *s*. : 7 *d*.——In general; it should be remembered, that 2 shillings allowed for every unit in the first place of decimals, and two-pence half-penny for every unit in the second place of decimals, will give, nearly enough, the value of the decimal part of every such expression.

(*b*) The value of two joint lives of 30, taken from Table XI. is 9.6. This subtracted from the value of the life in expectation, or from 13.1, by Table X. gives 3.5,

the

If either the rate of interest is supposed lower, or wives are supposed younger than their husbands, the annual payments will be increased. But there is no occasion for pointing out particularly the difference. It may be easily found in any cases by the directions now given. There is, however, one observation which ought to be here carefully attended to.—This method of calculation supposes, that the first annual payment is not to be made 'till the end of a year. If it is to be made *immediately*, the value of the joint lives will be increased one year's purchase; and, therefore, in order to find in this case the annual payments required, the value in present money found by Quest. I. must be divided by the value of the joint lives increased by unity, and, in this way, the preceding values at 4 *per cent.* according to the *Breslaw* Observations, will be found to be *l.*8.62—*l.*9.35—*l.*10.07.—According to the *London* Observations, *l.*10,—*l.*11.2,—*l.*12.7.

the number of years purchase which an annuity for a life of 30 years of age, *after* another life of the same age, is worth. This remainder, multiplied by 30, gives 105 *l.* the value in a single payment, supposing the reversionary annuity to be 30 *l.* And 105 *l.* divided by 9.6, gives *l.*10.9, the value of the same annuity in annual payments, during the joint continuance of the two lives, according to the *London* observations.—By similar operations all the other values above given have been found.

QUES-

Question V.

"A society may chuse to make abatements in these annual payments, and to require the remainder of the value of the reversionary annuity to be given, in fines or premiums at the time of admission. It may, for instance, chuse to fix the annual payments of all the members to 5 guineas. What, in this case, would be the premium due at admission, the annuity being supposed 30 *l. per annum*, and interest being at 4 *per cent?*"

Answer.

From the whole present value of the annuity in one payment, subtract the value of 5 guineas *per annum*, during the joint lives; and the remainder will be the answer.

Supposing the joint lives, both 40, the whole present value of the annuity in one payment is, according to the *Breslaw* Observations, *l.*101.1, by Quest. I.—The value of 5 guineas *per annum*, or of *l.*5.25 *per annum*, during two such joint lives, is *l.*5.25, multiplied by the value of the joint lives; that is, 5.25, multiplied by 9.82, or *l.*51.55; and this subtracted from *l.*101.1, gives *l.*49.5, the answer required for two lives at the age of 40.—The answer found in the same way for two lives whose common age is 30, is *l.*46.5,—and for two lives at 50, 50 *l.*

Accord-

According to the *London* Obfervations, thefe values are, for two lives at 30, *l*.54.6.—At 40, *l*.59.4.—At 50, *l*.63.3.

If the firft of the annual payments is to be made immediately; the true anfwer will, in every inftance, be the values found in the manner now directed, diminifhed by the annual payment; or, in the prefent cafe, 5 guineas lefs than the values fpecified.

The values, in *premiums* and *annual payments*, of any other reverfionary annuity, will be as much greater or lefs than thefe, as the annuity itfelf is greater or lefs.

QUESTION VI.

" A perfon 35 years of age wants to buy " an annuity, for what may happen to re- " main of his life after 50 years of age. " What is the value of fuch an annuity in " *ready money*, and alfo in *annual payments*, " 'till he attains to the faid age; that is, in " annual payments for 15 years, fubject in " the mean time to failure, fhould his life " fail?

ANSWER.

The prefent value of fuch an annuity is the *prefent* value of a life at 50, in money to be received 15 years hence, and the payment of which depends on the contingency of the continuance of the given life 15 years. That is, it is equal to the value of a life at 50, multiplied by the present value of 1 *l.* to be received at the end of 15 years, and also by the probability that the given life will continue so long.—A life at 50, according to Mr. *De Moivre*'s valuation of lives, and reckoning interest at 4 *per cent.* is worth 11.34 years purchase. The present value of 1 *l.* to be received at the end of 15 years, is, by Table I, 0.5553. And the probability that a life at 35, will continue 15 years, is, according to the *Breslaw* Observations $\frac{346}{490}$ *(a).* And these three values, multiplied by one another, give *l.*4.44, or the number of years purchase that ought to be given for the annuity.—The annuity then being supposed 50 *l.* its value in present money is 222 *l.*

(*a*) The probability that a given life shall continue any number of years, or attain to a *given age*, is (as is well known) the fraction, whose *numerator* is the number of the living in any Table of Observations opposite to the *given age* and *denominator*, the number opposite to the present age of the given life.—Thus, in the present instance; 346 is the number in Dr. *Halley*'s Table opposite to 50, and 490 the number opposite to 35.—$\frac{346}{490}$. (or the odds of 17 to 7) is, therefore, the probability that a person whose age is 35 shall attain to 50, or live 15 years. In the same manner it will appear, that, according to the same Table, the probability that a person at this age shall live 25 years, is $\frac{242}{490}$; or nearly an even chance.

At *Northampton* and *Norwich* a person at the same age, has an even chance of living 26 years; but in *London*, scarcely 20 years. See Tables III, IV, V, and VIII. Appendix. I will add, though foreign to my present purpose, that a person at the same age has in these towns a better chance of living one year, than in *London*, in the proportion of 3 to 2.

In

In order to find this value in *annual payments*, while the given life is attaining to 50, it is necessary to find the value of an annuity for 15 years, subject to failure on the extinction of the given life. And the value of such an annuity is, evidently, the last value subtracted from the value of the given life; or, in the present instance, *l*.4.44, subtracted from *l*.13.97. (See Table VI, Appendix) that is, *l*. 9.53.—222*l*. then, being the present value of an annuity of 50 *l*. for the remainder of a life now 35, after attaining to 50; and 9.53 being the number of years purchase, which ought to be given for an annual payment to last 15 years, if a life now 35 lasts so long, it follows, that the value of the same annuity in annual payments, 'till this life attains to 50, is 222 *l*. divided by 9.53; or *l*. 23:3.

This calculation supposes, that the first of the annual payments is not to be made 'till the end of a year. If the first payment is made immediately, the value will be, the *single payment* divided by the value of the life for the given term increased by unity; that is, in the present case, 222 *l*. divided by 10.53; or *l*.21.08.

If the value of the annuity is required in a single payment, over and above any given annual payment; deduct the value of the annual payment from the whole value in a single present payment, and the remainder will be be the anſwer.—Thus; let 5 guineas, in the preſent inſtance, be the given annual payment for the aſſigned term; and let the enquiry be, how much more in preſent money the ſuppoſed annuity is worth. By what has been juſt ſaid, 9.53, multiplied by 5 guineas, that is, 50 *l.* is the value of the annual payment; and this ſum deducted from 222 *l.* leaves 172 *l.* the anſwer.

If the annual payment begins immediately, its value is 10.53, multiplied by 5 guineas, and the anſwer comes out *l.* 166.75.

In this way may be found the value, in ſingle and annual payments, of any other annuity, payable to an aſſigned life, after a given term of years, taking any valuation of lives or intereſt of money. But care muſt be taken to remember, that it is the title to the annuity that will commence at the end of the given term, and that the firſt payment is not to be made 'till a year afterwards; that is, in the caſe here ſpecified, not 'till the end of 16 years.

Scholium.

The value of the *remainder* of two joint lives, after a given term of years, is likewiſe the value of 1 *l.* due at the end of the given term, multiplied by the value of two joint lives, each older by the given term than the given lives; and this product, multiplied by the probability, that the given joint lives ſhall

not

not fail in the given term; or (which is the same) by the product of the two probabilities, that the single lives shall each continue the given term. And the value of an annuity, on any given joint lives for a term of years beginning now, is this last value, subtracted from the whole present value of the joint lives. Thus; the value of two joint lives, one 40 years of age, and the other 50, (see Table VII.) is 8.91; which, multiplied by 0.6755, the value of 1 *l.* due 10 years hence, and by $\frac{445}{555}$, (the probability that a life at 30 shall continue 10 years) and also by $\frac{346}{445}$, (the probability that a life at 40 shall continue 10 years) gives 3.92, the present value of the remainder of two joint lives, aged 30 and 40, after 10 years; and this value, subtracted from 10.43, (the value in Table VII. of two joint lives, aged 30 and 40) leaves 6.51, their value for 10 years.

As the value of the longest of two lives is always the value of the *joint* lives, subtracted from the sum of the values of the two *single* lives; their value also for any *given term*, is the value of the *joint* lives for the given term, subtracted from the sum of the values of the *single* lives for the given term.

The truth of these rules may easily appear without particular proof. I have, however, pointed out the method of demonstrating them in a note *(a)* at the end of this work.

(*a*) See note (B) in the Appendix.

By similar operations, may be found the values of 3 or more *joint* lives, or the longest of *three* or more lives, for a given term of years, or of what shall remain of them after a given term of years.

Question VII.

" The present value is required of an annuity to be enjoyed by one life, for what may happen to remain of it beyond another life, after a given term; that is, provided *both* lives continue, from the present time, to the end of a given term of years?"

Answer.

Find the value of the annuity for two lives greater, by the given term of years, than the given lives. Discount this value for the given term; and then, multiply by the probability, that the two given lives shall *both* continue the given term; and the product will be the answer.

Example.

Let the two lives be each 30. The term seven years. The annuity 10*l*. Interest, 4 *per cent.*——The given lives, increased by 7 years, become each 37. The value of two joint lives each 37, is (by Table VII.) 10.25.

The

The value of a single life at 37, is (by Table VI.) 13.67. The former, subtracted from the latter, is 3.42, or the value of an annuity for the life of a person 37 years of age, after another of the same age, by Quest. I.—3.42 discounted for 7 years, (that is, multiplied by 0.76, the value of 1 *l.* due at the end of seven years, by Table I.) is 2.6.—The probability that a single life at 30 shall continue 7 years, is (by the hypothesis explained page 2.) $\frac{49}{56}$ *(a)*. The probability, therefore, that two such

---

(a) In this case, it is, on some accounts best, as well as easiest, to take the probabilities of life from the hypothesis, rather than immediately from the Tables.—Fifty-six persons being supposed alive at 30, one will die every year, according to the hypothesis. At the end of seven years then, the number of the living will be 49, and $\frac{49}{56}$, or the odds of 7 to 1, is, by note, p. 18, the probability, that a life, aged 30, will continue 7 years; and this fraction, multiplied by itself, is the probability, that two lives of this age, shall *both* continue 7 years. In general, it must be remembered, that the probability, that any two or more events shall *all* happen, is the product arising from multiplying by one another, the probabilities of all the events taken separately. The probability, therefore, that any number of persons will *all* live any given time, is rightly found by multiplying into one another the probabilities that each of them will live that time.—It may further be of use to some, that I should observe here, that the difference between unity and the fraction expressing the probability, that an event will happen, gives the probability that it will *not* happen. Thus; the probability, that a person 40 years of age will live 11 years, is, by the *Breslaw* Table $\frac{225}{445}$. The probability, therefore, that he will *not* live 11 years, is $\frac{225}{445}$, subtracted from

such lives shall both continue 7 years, is $\frac{2401}{3136}$, or, in decimals 0.765. And 2.6, multiplied by 0.765, is 1.989, the number of years purchase which ought to be given for an annuity, to be enjoyed by a life now 30 years of age, after a life of the same age, provided both continue 7 years. The annuity then being 10*l.* its present value is *l.* 19.89.

By similar operations, it may be found, that supposing the term one year, and the ages and the rate of interest the same, the present value of the same reversionary annuity is *l.* 32.4; and that if the term is 15 years, the value is *l.* 9.7.

For two lives each 40, these values are *l.* 30.33.—*l.* 17.44.—*l.* 7.3. the term being 1, 7, or 15 years.

For two lives each 50, the same values for the same terms, are *l.* 28.2,—*l.* 13.86,—*l.* 4.34 *(a)*.

These values, according to the *London* Observations and Mr. *Simpson*'s Tables of the values of single and joint lives, are,

from unity or $\frac{110}{443}$.—In like manner: The probability that two persons aged 30, shall *both* live 7 years, being 0.765, the probability that they will *not* both live so long, or that *one or other* of them will die in 7 years, is 0.765, subtracted from unity, or .235.

If any reader is unwilling to take these assertions for granted, he should consult the beginning of Mr. *De Moivre*'s, or Mr. *Simpson*'s Treatises on the Doctrine of Chances, where he will find them demonstrated.

(a) See Note (C) Appendix.

For 2 lives at 30—*l.*32.05—*l.*18.62—*l.*7.66.
at 40—*l.*30.7 —*l.*15.6 —*l.*5.45.
at 50—*l.*29.36—*l.*12.33—*l.*3.24.

QUESTION VIII.

"Let the scheme of a society for granting annuities to widows, be, that if a member lives *a year* after admission, his widow shall be entitled to a life annuity of 20 *l.* If *seven* years, to 10 *l,* more, or 30 *l.* in the whole. If *fifteen* years, to another additional 10 *l.* or 40 *l.* in the whole. What ought to be the annual payments of the members for the ages of 30, 40, and 50, supposing them of the same ages with their wives, and allowing compound interest at 4 *per cent.* ?

ANSWER.

According to the *hypothesis*, explained p. 2; and, therefore, very nearly, according to the Tables of Observation for *Breslaw*, *Norwich*, and *Northampton*,

l. 8.44—*l.* 8.69—*l.* 9.05.

According to the *London* Observations,

l. 9.41—*l.* 10.17—*l.* 10.92.

These

These values are easily deduced from the values in the last Question. For example. The value of 10 *l. per annum* for life to 40 after 40, provided the joint lives do not fail in *one* year, is, according to the *hypothesis*, *l*.30.33. The value of 20 *l. per annum*, in the same circumstances, is, therefore, *l*.60.66.— In like manner, the value of 10 *l*. after *seven* years, is *l*. 17.44. And of 10 *l*. after 15 years *l*.7.3.—These values together make *l*.85.4, or the value of the expectation, described in this Question, in a *single present payment*; which, divided by 9.82, (the value by Table VII. of two joint lives at 40) gives *l*.8.69, the value of the same expectation in *annual payments*, during the joint lives.—In the same manner may be found the answer in all cases to any Questions of this kind.

These calculations suppose, that the annual payments do not begin 'till the end of a year. If they are to begin *immediately*, the true *annual payments* will be, as was before observed, the *single* payments, divided by the value of the joint lives increased by unity; and in the present case they will be, by the *hypothesis*,

l. 7.75—*l*. 7.9—*l*. 8.07.

By the *London* Observations,

l. 8.52—*l*. 9.06—*l*. 9.51.

By

By the method of calculation now explained, may be easily found in all cases, supposing the annual payments previously settled, what the reversionary annuities are corresponding to them in value.—Thus, the annuities being the same with those mentioned in this Question, the *mean* annual payments for all ages between 30 and 50, are nearly 8 *l.* according to the *highest* probabilities of life; 9 *l.* according to the *lowest*; and 8 guineas the *medium (a)*; interest being at 4 *per cent.* and the first payment to be made immediately.

If the mean annual payments, beginning immediately, are fixed to five guineas, the corresponding life annuities will be nearly (by the *hypothesis*) 12 *l.* if the contributor lives a year, and 24 *l.* if he lives seven years; or (by the *London* Observations) 12 *l.* if he lives a year, and 20 *l.* if he lives seven years *(b)*.

It

(*a*) The value of this expectation, supposing married men 40 years of age, and their wives 30, is, in a *single* payment, 113 *l.* In annual payments, beginning immediately, *l.* 9.88, by the *hypothesis*. And 107 *l.*—and *l.* 10.93, by the *London* Observations.

(*b*) If the annuities in expectation are 14 *l.* provided a member lives a year, and 20 *l.* provided he lives seven years, the proper *mean single* payments for all ages, taken one with another, under 50 or 52, is 50 guineas nearly, according to all the Tables of Observation, supposing equality of age between men and their wives. And the addition which ought to be made, on account of excess of age on the man's side is, taking the nearest and the

easiest

It is obfervable, that the difference in the values of the annuities, arifing from difference of ages, and the difference in the probabilities of life, is lefs in this Queftion than in Queftion 4th; and that, confequently, the plan propofed in it, is the fafeft, as well as the moft equitable and encouraging, that a fociety can adopt.

It is neceffary to remark here further, that *yearly* payments which begin immediately, are more advantageous than *half-yearly* payments which begin immediately. Mr. *Simpfon* (in his Treatife on *The Doctrine of Annuities and Reverfions*, p. 78, and alfo in his *Select Exercifes*, p. 283.) has fhewn, that, in the cafe of life annuities, *half-yearly* payments, which begin at the end of half a year, are $\frac{1}{4}$ of a year's purchafe better than *yearly* payments, which begin at the end of a year. And it is manifeft, that *half-yearly* payments, which begin immediately, are no

---

eafieft round fums, about a guinea and $\frac{1}{2}$ for every year as far as 17 years; or, in the annual payments, (fuppofed 5 guineas) $\frac{1}{4}$ a guinea *per annum* for five years excefs, and $\frac{1}{4}$ a guinea more for every four years excefs beyond five years, 'till the excefs comes to be 17 years. And, I believe, that 60 guineas in *fingle payments*, and fix guineas in *annual payments* beginning immediately, may very well be ftated as the *loweft common* payments proper to be required, fuppofing all married men under 52, taken into a fociety, without enquiring into the difference of age between them and their wives, the annuities being all along fuppofed to be *life* annuities, and intereft reckoned at 4 *per cent*.

more

Reversionary Annuities, &c.   29

more than half a year's purchase better than those which begin at the end of half a year. But *yearly* payments, which begin immediately, are a *whole year*'s purchase better than the same payments to begin at the end of a year. The difference of value, therefore, between *yearly* and *half-yearly* payments, supposing both to begin immediately, is a quarter of a year's purchase in favour of the former.

QUESTION IX.

" The value is required of an annuity to
" be enjoyed for what may happen to re-
" main of one life after another, provided
" the life in expectation continues a given
" time?"

ANSWER.

Find by Question VI. the present value of the annuity for the remainder of the life in expectation, after the given time, and multiply this value by the probability, that the other life shall fail within that time. Find also, by Question VII, the value of the reversion, provided *both* lives continue the given time. Add these values to one another, and the *sum* will be the answer in a single present payment.

EXAMPLE.

An annuity of 10 *l.* for the life of a person now 30, is to commence at the end of 11 years *(a)*, if another person now 40, should be then dead; or, if this should not happen, at the end of any year beyond 11 years in which the former shall happen to survive the latter. What is the present value of such an annuity, reckoning interest at 4 *per cent.* and taking the probabilities of life as they are in Dr. *Halley*'s Table?

The value of 10 *l. per annum*, for the remainder of the life of a person now 30, after 11 years, found by Quest. VI. is *l.*69.43.—The probability that a person 40 years of age shall live 11 years, is, by Dr. *Halley*'s Table, $\frac{311}{447}$. The probability, therefore, that he will die in 11 years, is $\frac{311}{447}$ subtracted from unity *(b)*, or $\frac{110}{447}$; which multiplied by *l.*69.43, gives *l.*17.16.—The value of the reversion, provided *both* live 11 years, found by Quest. VII. is 17 *l.* And this value added to the

---

(a) That is, the title to the annuity is to commence at the end of 11 years, and the first payment to be made a year afterwards, in case the life in expectation should continue so long, and the other fail. But if *both* lives should continue the given term, the first payment is always to be made at the end of the year, in which the former life shall happen to survive the latter. See Quest. VI.

(b) See the Note, p. 23.

former, makes *l.* 34.16, the value required in a *single present payment*; which payment divided by *l.*11.43, (the value by Table VII. of two joint lives, aged 30 and 40, with unity added) gives 3 *l.* *(a)*; or the value required in annual payments during the joint lives, the first payment to be made immediately.—If, every thing else being the same, the assigned term is 15 years, the value required will be 29 *l.* in a *single payment*, and *l.* 2.55 in *annual payments*.

QUESTION X.

"What money in hand, and also in an-
"nual payments during life, ought a person
"of an assigned age to give for a sum of mo-
"ney, payable at his death to his heirs *(b)*?—
"In other words, what money in hand, and
"in annual payments during life, ought a
"person of a given age to pay for an *assu-
"rance* of any given sum on his life?"

ANSWER.

Subtract the value of the life from the *perpetuity*. Multiply the remainder by the

---

(a) See the demonstration of this rule in Note (D) Appendix.

(b) This Question is the same with Problem 16th, in Mr. *De Moivre*'s Treatise on Annuities, and Problem 26th, in Mr. *Simpson*'s Select Exercises; but the answers there given are right only when applied to reversionary *estates*, and therefore must be materially wrong, when applied to reversionary *sums*, as will appear from the *Scholium* to this Question, and from note (E) in the Appendix.

product

product of the given fum into the intereft of 100 *l.* for a year: and this laft product, divided by 100 *l.* increafed by its intereft for a year, will give the anfwer in a *fingle prefent* payment. And this payment, divided by the value of the life, will give the anfwer in *annual* payments, during the continuance of the life.

Example. Let the life be 30. The fum 100 *l.* The rate of intereft 4 *per cent*. And the valuation of lives, that in Table VI. The perpetuity, therefore (*a*), is 25. The intereft of 100 *l.* for a year, is 4 *l.* 100 *l.* increafed by its intereft for a year, is 104 *l.* And the value of the life 14.68.—The value of the life, fubtracted from the perpetuity, gives 10.32, which, multiplied by the product of 100 *l.* into 4, or by 400, gives 4128. And this, divided by 104, gives *l.* 39.7, the value of 100 *l.* payable at the death of a perfon aged 30, in a fingle prefent payment.—And this payment, divided by 14.68, is *l.* 2.7, the fame value in annual payments during the continuance of the life.

Thefe values found in the fame way agreeably to the valuation of lives for *London*, in Table X, are *l.* 45.76, and *l.* 3.49.—If the life is 36, and intereft 4 *per cent*. thefe values are 43 *l.* and *l.* 3.1, by Table VI; and *l.* 49.6,

---

(*a*) That is; the value of the *fee-fimple* of an eftate found by dividing 100 *l.* by the rate of intereft.

and *l*. 4.1, by Table X.—If interest is reckoned at 3 *per cent.* the same values are, by Table VI, for 30 years of age, *l*. 48.14.— 2.86.—For 36 years of age, *l*. 51.43, and *l*. 3.28.

It appears here, that difference of interest makes no considerable difference in the answers to Questions of this kind, except when the values are required in a single payment.

If the first of the annual payments is to be made immediately, the single payment is to be divided by the value of the life, with unity added to it, agreeably to what has been already observed; and the annual payments in this case (interest supposed at 4 *per cent.*) will be by Table VI, for a life at 30, *l*. 2.53— At 36, *l*. 2.9.

If the payments are half-yearly payments beginning immediately, the single payment must be divided by the value of the life increased by ¼, or .75, (see Quest. VIII.) And the half-yearly payments, for the age of 36, will be half 2.9, or 1.45. And half 1.45, or .725, is likewise nearly the proper quarterly payments.

Again; if an annual payment, beginning immediately, of *l*. 2.9, ought (reckoning interest at 4 *per cent.*) to purchase 100 *l*. payable at the failure of a life now 36; 5 *l*. by the rule of proportion, ought to purchase 172 *l*. And in like manner, it may be found, that the same annual contribution, in half-

D yearly

yearly or quarterly payments, beginning immediately, ought to purchafe 170 *l.*—Thefe fums, according to the *London* Obfervations, are 132 *l.* and 130 *l.* nearly.

The reafon of mentioning thefe particulars will be feen in the next chapter.

Scholium.

If the reverfion is not a *fum*, but an annuity for ever, or an *eftate* in *fee-fimple*, to be entered upon after a given life, its prefent value, *in a fingle payment*, will be " the value " of the life fubtracted from the perpetuity, " and the remainder multiplied by the an- " nuity, or the annual rent of the eftate."— And the value, in *annual payments*, will be, as before, the fingle payment divided by the value of the life.—Univerfally. It ought to be remembered, that a reverfionary *eftate*, after any given life or lives, is worth as much more than a correfponding reverfionary *fum*, as 100 *l.* increafed by its intereft for a year, is greater than 100 *l.*—Thus, the prefent values, in fingle and annual payments, of 4 *l.* *per annum* for ever, and of 100 *l.* in money after any affigned life, are to one another, (intereft being at 4 *per cent*.) as 104 to 100, or 1.04 to 1.—The reafon of this difference is, that the calculations fuppofe, that the reverfionary *fum*, and the firft yearly rent of the *eftate*, or firft payment of the annuity,

are

are to be received at the same time, after the extinction of the lives in possession. It is easy to see, that this is a circumstance which must make the latter of most value. But to prevent any doubts about it, I shall explain it more particularly in a note in the Appendix *(a)*.

Question XI.

" A person of a given age, having a year-
" ly income which will fail with his life,
" wants to make provision for another per-
" son of a given age, in case the latter should
" happen to survive. What ought the for-
" mer to give in a single payment, and also
" in annual payments during their joint lives,
" for a given sum, payable at his death to
" the latter ?

It is manifest, that the value of the given sum in this case, must be less than in the case stated in the last Question; because, here the payment of it is suspended on the contingency, that one life shall survive another; whereas in the other case, it is *certainly* to be paid at the failure of a given life.

Answer.

Find, by the solution of problem 32d, p. 297, Mr. *Simpson*'s Select Exercises, the

(a) Vid. Appendix, note (E).

value of an estate, corresponding to the given sum, and depending on the given survivorship. Divide this value by 1 *l.* increased by its interest for a year, and the quotient will be the value of the given sum in a single present payment. And the single payment, divided by the value of the given joint lives, will be the answer in annual payments during the joint lives.

The solution I have referred to is as follows.

" Find the value of an annuity on two
" equal joint lives, whereof the common age
" is equal to the age of the older of the two
" proposed lives; which value, subtract from
" the perpetuity, and take half the remain-
" der. Then say, as the *expectation* of the
" duration of the younger of the two lives is
" to that of the elder, so is the said half re-
" mainder to a 4th proportional, which will
" be the number of years purchase to be gi-
" ven for the estate when the life in expec-
" tation is the oldest of the two. But if this
" life is the youngest, then add the number
" of years purchase just found to the value
" of the joint lives, and let the sum be sub-
" tracted from the perpetuity, and you will
" also have the answer in this case *(a).*"

Let

(*a*). Mr. *Simpson* has given the following examples of this solution, adapted to *London* lives.——Example I. " Suppose the age of the *expectant* to be 40; of the *pos-* " *sessor* 30. The rate of interest 4 *per cent.* and the
" given

Let the life in expectation be 30; and the other life 40: The sum, 100 *l*. Interest, 4 *per cent*. The valuation of lives, that in Table VI.

The *expectation* of the first life, is 28; of the second life 23, by Mr. *De Moivre's hypothesis*. The value of the joint lives is 10.43,

" given legacy 5000 *l*. or 200 *l. per annum*. Then the
" value of two equal joint lives of 40, being 8.1, by
" Table XI, and the perpetuity 25, the remainder or
" difference will be here 16.9; whereof the half is 8.45.
" Therefore, it will be as 23.6 to 19.6, so 8.45 to 7.02
" years purchase, or *l*. 1404, the required value."

Example II. " Let the age of the *expectant* be 30, of
" the *possessor* 40, and the rest as in the preceding exam-
" ple. Here the value of the joint lives 30 and 40, will
" be 8.8; which added to 7.02, (found above) the sum
" will be 15.82; whence the answer, in this case, is
" 9.18 years purchase, or 1836."

I have shewn, that the values of reversionary *estates*, and reversionary *sums*, are not the same as is here supposed.—The rule gives the true value when applied to the former; but, when applied to the latter, the values given by it must be divided by 1 *l*. increased by its interest for a year, as above directed.—The same observation is to be applied to Mr. *Simpson's* next Problem, or the 33d.

In these Examples 23.6 and 19.6, are the expectations, in Table IX, of 30 and 40, according to the *London* Tables of Observation; and the method of finding them for any age, and from any Tables of Observation, is explained at the beginning of the first Essay.

In Mr. *De Moivre's hypothesis*, the expectation of a life, is always *half* the complement. See note, p. 2.—Sometimes the *complement* of a life is mentioned without any view to Mr. *De Moivre's* hypothesis, and it then means double the *expectation* of the life, whatever that may be, according to any Table of Observations.

by Table VII. The value of two joint lives, both 40, is 9.82, by the same Table. The estate corresponding to 100 *l.* is 4 *l. per. ann.* and the present value of such an estate to be entered upon by a person 30 years of age, provided he survives a person 40 years of age, is, by the rule just quoted, *l.* 33.32. And this value, divided by 1 *l.* increased by its interest for a year, or by 1.04, is *l.* 32.03. the value in a *single present payment* of the sum of 100 *l*, dependent on the given survivorship. And this single payment, divided by 10.43, is *l.* 3.07, the required value in *annual payments*, during the joint lives, if the first payment is not to be made 'till the end of a year. But if the first payment is to be made immediately, the required value in *annual payments* will be *l.* 32.03, divided by 11.43, or *l.* 2.8.—These values, according to the *London* Observations, or Mr. *Simpson*'s Tables founded upon them, are *l.* 35.30, in a *single payment*, and *l.* 3.6, in *annual payments*, beginning immediately.

Mr. *Simpson*, in the Problems following that here quoted, has given solutions of most other Questions, concerning the values of reversions depending on survivorships, where the whole duration of two or three lives is concerned. And I am acquainted with no other solutions of these Questions, which are applicable to all Tables of Observations, and which at the same time (proper regard being paid

paid to the correction explained in the last Question) may be considered as sufficiently correct *(a)*.

QUESTION XII.

" Suppose an institution for the relief of widows to extend its assistance likewise to the families of married men, provided they leave no widows. Suppose, for instance, that in this case children are to be entitled to 100 *l*. What is such an expectation worth, in present payment, according to Dr. *Halley*'s Table, interest being at 4 *per cent.* ?"

ANSWER.

If 40 is the mean age at which members are admitted on such an institution, and 32 the mean age of their wives, the answer (supposing no subsequent marriages) is, by the 33d Problem in Mr. *Simpson*'s Select Exercises, p. 298, and the correction already explained, *l*. 13.80 *(b)*.

But

(a) See the third Essay.

(b) This Problem and its solution are given by Mr. *Simpson* in the following words: " A and his heirs are entitled to an estate of a given value, upon the decease of B, provided B survives A; to find the value of their expectation in *present* money."—Solution. " Find the value of an annuity on the longest of two equal " lives,

But there is a reduction necessary, on account of the chance there is, that a widower may marry again. Suppose, therefore, one half of all widowers to marry a second and third time, and that two-fifths of such widowers survive these subsequent marriages. In this case, $\frac{1}{2}$ added to $\frac{2}{5}$ of $\frac{1}{2}$, or $\frac{7}{10}$ of all who become widowers, will die without leaving widows, and therefore $\frac{7}{10}$ of $l. 13.8$, or $l. 9.66$, will be the answer. If only *one fourth* of all who become widowers marry again, and two fifths of these survive, the answer will be $l. 11.73$.

" lives, whereof the common age is that of the older of the lives A and B; which value subtract from the perpetuity, and take half the remainder; then it will be, as the expectation of duration of the younger of the lives A and B, is to that of the older, so is the said half remainder to the number of years purchase required, when the life of B *is the older of the two*. But if B *be the younger*; then to the number thus found, add the value of an annuity on the longest of the lives A and B, and subtract the sum from the perpetuity, for the answer in this case."

If the estate is $4 l.$ *per annum*, the age of B 40, and of A 32, interest 4 *per cent.* the answer by this rule comes out $l. 14.35$, which divided (as in the preceding Question) by 104, gives $l. 13.80$, the value, as above, of $100 l.$ in money. If B is 30 and A 40, the same value is $20 l.$

N. B. The value of the longest of two lives is always the *difference* between the value of the *joint* lives, and the *sum* of the values of the two given *single* lives. Thus; the value of a life at 40, is, by Table VI, 13.2. The *sum* of the values of two such lives, is 26.4. The value of two joint lives, whose common age is 40, is, by Table VII, 9.82; and the difference is 16.58, or the value of the *longest* of two lives at 40.

This

This calculation supposes all marriages to leave children who survive their parents. If this is considered as uncertain, the values now determined must be diminished in the proportion of this uncertainty.—Thus; if one marriage in seven fails of leaving children (*a*) that survive their parents; these values will be reduced a *seventh* part, or to *l*.8.28, if *half*, and *l*. 10.05, if a *quarter* of all widowers marry.

In this way may any other questions of the same kind be answered on any suppositions that may be thought most reasonable.

Question XIII.

" Let an establishment be supposed which
" takes in at once all the marriages in a
" country, or all marriages among persons
" of a particular profession within a given
" district, and subjects them for perpetuity
" to a certain equal and common tax, or an-
" nual payments, in order to provide life an-
" nuities for such widows as shall result from
" these marriages. What ought the tax to
" be, supposing the annuity 20 *l.* and calcu-
" lating at 4 *per cent*. from Mr. *De Moivre*'s
" valuation of lives; or, which is nearly the
" same, from the probabilities of life in Dr.
" *Halley*'s Table of Observations?"

(*a*) This for many years has been nearly the fact among the ministers and professors in *Scotland*.

Answer.

Answer.

Since at the commencement of such an establishment, all the oldest, as well as the youngest marriages, are to be entitled equally to the proposed benefit, a much greater number of annuitants will come immediately upon it, than would come upon any similar establishment, which limited itself in the admission of members to persons not exceeding a given age. This will check that accumulation of money, which should take place at first, in order to produce an income equal to the disbursements at the time when the number of annuitants comes to a *maximum*; and, therefore, will be a particular burden upon the establishment in its infancy. For this, some compensation must be provided; and the equitable method of providing it, is, by levying *fines* at the beginning of the establishment, on every member *exceeding* a given age, proportioned to the number of years which he has lived beyond that age. But in the present question, it is supposed, that such fines cannot be conveniently levied, or that every payment must be equal and common, whatever disparity there may be in the value of the expectations of different members. The fines, therefore, must be reduced to one common one, answering as nearly as possible to the disadvantage I have mentioned, and
payable

payable by every member at the time when the establishment begins. After this, the establishment will be the same with one that takes upon it all at the time they marry; and the tax or annual payment of every member adequate to its support, will be the annual payment during marriage, due from persons who marry at the mean age at which, upon an average, all marriages may be considered as commencing.—There are then two points to be here determined. The *fines* necessary to be paid at first, according to the account I have just given; and the *constant annual payment*, necessary to be made by every member, as an equivalent for the expectation provided by the establishment.—The *fines* to be paid at first are, for every particular member, the same with the difference between the value of the expectation to him at his present age, and what would have been its value to him had the scheme begun at the time he married? Or, they are, for the whole body of members, the difference between the value of the common expectation, to persons at the mean age of all married persons taken together as they exist in the world, and to persons at that age, which is to be deemed their mean age when they marry.

Thus; let 33 for the man, and 25 for the woman, be the mean ages of all that marry annually. Let also 48 be the mean age of all the married men in the world, and 40 of

married

married women *(a)*.—Now, he that will calculate for these ages, in the manner directed in Quest. IV. will find, that the value in *annual payments* during marriage, and beginning immediately, of the expectation of an annuity of 20 *l. per annum* by a person 25 years of age, after a life whose age is 33, is *l.* 6.64.—And that *l.* 8.04, is the value of the same expectation, the ages being 48 and 40.

The former, therefore, is the payment for perpetuity from every member of the establishment; and the value of the *difference* between it and the latter, or of *l.* 1.4 *per ann.* payable during two joint lives, whose ages are 40 and 48, that is, *l.* 14.2, is the fine necessary to be levied on every married member at the beginning of the establishment *(b)*.

It would be easy to extend the benefit of such an establishment, so far as to provide 100 *l.* for the children of members, provided

---

(a) I must beg leave to refer to note (F) in the Appendix, for an explanation of what I mean by the mean ages of married men and women, and also for a confirmation of the answer I have given to this Question.

(b) An annuity for ever, the first payment of which is to be made immediately, is worth 26 years purchase, interest being at 4 *per cent.* *l.* 14.2 therefore, is equivalent in value to 0.55 *l.* or 11 *s. per annum*, for ever. Add this to *l.* 6.64, and it will appear, that *l.* 7.19 *per annum*, beginning immediately, is the answer to this Question, supposing the value of the *fine* to be provided for in the perpetual annual payments.

they

they leave no widows; and the necessary addition on this account to the perpetual annual payments, can scarcely, in the circumstances this question supposes, be much more than about 15 *s.* payable during life, and excluding from all benefit such as happen to be widowers at the commencement of the establishment, and do not afterwards marry.

If, in such an establishment, all persons of a particular denomination, whether married men, widowers, or batchelors, are subjected alike to the taxes and fines; they ought to be as much *less,* as the whole number of persons subjected to them, is *greater* than the number of marriages constantly existing.

In carrying these schemes into execution, there cannot be a more easy, or equitable way of raising the necessary fines, than by providing, that none shall be entitled to any expectation for a few of the first years. Thus; an establishment, entitling widows to 20 *l. per annum* for life, and consisting of 667 married members, and 344 unmarried, always kept up at an average, ought to begin with a capital of *l.* 14.2 multiplied by 667, or 9471 *l.* besides one payment in hand of the constant annual payments. That is, (the proper annual payment of every member being in this case $\frac{667}{1011}$, multiplied by *l.* 6.64, or *l.* 4.38) it ought to begin with a capital of 13,899 *l.* over and above the payment of *l.* 4.38, at the *end* of every year for ever afterwards (*a*).—The exclusion of all the first members from any benefit, unless they survive the first *two* years, or live to make *three* payments, would raise this capital nearly. And such an exclusion for *three* or *four* years, would be an advantage so considerable, that it would probably give security and stability to the scheme for all subsequent time.

In these observations, I have had in view several schemes of the kind described in it; which are now actually established in this kingdom; but more particularly, one begun among the *London* and *Middlesex* clergy, and another which is established by act of parliament among the clergy in *Scotland*; of both which, I shall have occasion in the next chapter to take further notice.

I have chosen to calculate here only from Dr. *Halley*'s Table, or Mr. *De Moivre*'s *hypothesis* grounded upon it, because the *London* Table is, by no means, adapted to the cases in view.

It should be further remembered, that when the mean ages, at which marriages commence, are supposed to be 33 and 25,

---

(*a*) Or, supposing the value of 9471 *l.* (the fine) provided for in the annual payments, it ought to receive every year, at the *beginning* of the year, a contribution from each member of *l.* 4.74.

all second and third marriages are included; and that it is to be expected, that almost all these marriages will begin after these ages; and likewise, that a considerable proportion of the first marriages will begin a much longer time *after* these mean ages, than any of the other first marriages will begin before them.—Probably, therefore, these mean ages should not be taken younger. One or two years, however, more or less, in every supposition I have made, will make no difference of any consequence.

QUESTION XIV.

" A person of a given age has an estate depending on the continuance of his life for a given term. What ought he to give for having it *assured* to him for that term?"

ANSWER.

From the value of an annuity certain for the given term, found by Table II, subtract the value of the life for the given term, found by Quest. VI. and *reserve* the remainder.—Multiply the value of 1*l*. due at the end of the given term, (found by Table I.) by the *perpetuity*, and also by the *probability*, that the given life shall fail in the given term. The *product* added to the *reserved* remainder, and the *sum* multiplied by the given annuity,

will

will be the required value of the affurance in one prefent payment *(a)*.

EXAMPLE.

An eftate or annuity of 10 *l. for ever*, will be loft to the heirs of a perfon now 34, fhould his life fail in 11 years. What ought he to give for the *affurance* of it for this term?—That is; What is the prefent value of fuch an annuity to be entered upon at the failure of fuch a life, fhould that happen in 11 years?

The value of the life of a perfon whofe age is 34 for 11 years, is, by Queft. VI. (reckoning intereft at 4 *per cent.* and calculating from Dr. *Halley*'s Table of Obfervations) 7.76; which, fubtracted from 8.760, (the value of an anuuity certain for 11 years) leaves 1 *l.* the *remainder* to be referved.

The value of 1 *l.* to be received at the end of 11 years, is, 0.6496, by Table I. The probability that the life of a perfon, aged 34, fhall fail in 11 years, is, by Dr. *Halley*'s Table, $\frac{103}{499}$; and in the perpetuity is 25. Thefe numbers, multiplied by one another, and 1 added to the product, make 4.34, which, multiplied by 10, (the given annuity) gives *l.* 43.4, the required value in a fingle prefent payment.

(a) See the demonftration in note (G) Appendix.

l. 43.4s

l.43.4, divided by 1.04, gives *l*. 41.7, the true value, by Scholium to Queſt. X. of the aſſurance of an *equivalent ſum*, or of 250*l*. for 11 years on the given life.

Again. 41.7, divided by 8.76, (the value of the given life for the given time with unity added to it) gives 4.76, the ſame value in annual payments beginning immediately, for 11 years (*a*), ſubject to failure ſhould the life fail.

SCHOLIUM.

In a ſimilar way may the price of aſſurances on any two joint lives, or the *longeſt* of two lives for any given terms, be calculated; the rule being as follows:

" From the value of an annuity certain
" for the *given term*, ſubtract the value of
" the joint lives, or the longeſt of the two
" lives for the *given term*, found by Scho-
" lium to Queſt. VI. and *reſerve* the remain-
" der.—Multiply the value of 1 *l*. to be re-
" ceived at the end of the given term by the
" perpetuity, and alſo by the probability
" that the *joint lives*, or the *longeſt of the two*
" *lives*, ſhall fail within the given term. This
" product added to the reſerved *remainder*,
" and the *ſum* multiplied by the annuity to be

(*a*) The laſt payment to be made at the end of the 11th year; or 12 payments in all.

" aſſured,

" assured, will be the value of the assurance
" in a single present payment."

EXAMPLE.

" What is the value of 10 *l. per annum*, to
" be entered upon, should *either* of two
" persons, one 40 and the other 30 years of
" age, die in ten years, reckoning interest
" at 4 *per cent.* and calculating from Dr.
" *Halley*'s Table."

The value of two joint lives at these ages, for 10 years, (found by *Scholium* to Quest. VI.) is 6.51; which, subtracted from 8.111, (the value of an annuity certain for 10 years, at 4 *per cent.*) leaves 1.60, the remainder to be *reserved*.

The value of 1 *l.* to be received at the end of 10 years, is, .6755, by Table I.

The probability, that the lives of one or other of two persons, aged 30 and 40, shall fail in 10 years, is, $\frac{181}{571}$ by Table III. (*a*). And the perpetuity 25. These numbers, multiplied by one another, and 1.60 added to the product, make 7.48, which, multiplied

---

(*a*) The probability taken from the Table, that a person aged 30, shall live 10 years, is, $\frac{445}{473}$. That a person, aged 40, shall live 10 years, is, $\frac{346}{445}$. That they shall *both* live 10 years, is, $\frac{346}{445}$, multiplied by $\frac{445}{473}$, or $\frac{346}{571}$. That they shall *not both* live 10 years, or that *one* or *other* of them shall die in this time, is, $\frac{346}{571}$, subtracted from unity, or $\frac{185}{571}$. See note p. 23.

by

by 10, (the given annuity) gives *l.* 74.8, the anfwer in a fingle prefent payment.

l. 74.8, divided by 1.04, gives *l.* 71.92, the value of the affurance of an *equivalent fum*; or of 250 *l.*—*l.* 71.92, divided by 7.51, (the value of the two joint lives for 10 years with unity added) gives 9.57, the value of the fame fum in annual payments beginning immediately, for 10 years, fubject to failure fhould the joint lives fail.

Example II.

" What is the value of 10 *l. per ann.* to be
" entered upon, fhould two perfons, one 30,
" and the other 40, *both* die; that is, fhould
" the *longeft* of the two lives fail in 10 years,
" reckoning intereft at 4 *per cent.* and cal-
" culating from Dr. *Halley*'s Table?"

The value of the *longeft* of the two lives for 10 years, (that is, the value of the joint lives for 10 years, fubtracted from the fum of the (*a*) values of the fingle lives for 10 years) is, 7.91; which, fubtracted from 8.111, the value of an annuity certain for 10 years, leaves .20 the remainder to be referved.— The value of 1 *l.* to be received at the end of 10 years, is, .6755. The probability that the lives of two perfons, aged 30 and 40, fhall fail in 10 years, is, by Table III, $\frac{86}{171}$,

(*a*) See Scholium to Queft. VI.

multiplied by $\frac{29}{447}$, or $\frac{8514}{216191}$; and the perpetuity 25. These numbers, multiplied by one another, and .20 added to the product, make .740, which, multiplied by 10, (the given annuity) gives 7.4, the answer in a single payment.

7.4, divided by 1.04, gives 7.11, the value of the assurance of 250 *l*.

REMARK I.

The values of single lives for given terms, when these terms are less than ten years, must, in answering these Questions, and also in answering the following Questions, be found true to at least 2 or 3 places of decimals. But they cannot be found to this exactness by any Tables that are extant; and, therefore, they must be calculated in the following manner:

" Multiply the probability, taken out of
" the Table of Observations, that the life
" shall exist 1, 2, 3, &c. years, by the value
" of 1 *l*. due at the end of 1, 2, 3, &c. years;
" and the sum of the products will be the
" value of the life for 1, 2, 3, &c. years."

For Example. The probability, that a person whose age is 34, shall live a year, is, by Dr. *Halley*'s Table, $\frac{490}{499}$. The probability, at the same age, of living 2 years, is, $\frac{481}{499}$; 3 years, $\frac{472}{499}$.—$\frac{490}{499}$ multiplied by .9615, (the value,

value, by Table I. of 1 *l*. due at the end of a year, intereft being at 4 *per cent.*) is, .942; or the value of the life for *one* year.—$\frac{481}{499}$, multiplied by .9245, (the value of 1 *l*. due at the end of 2 years) is, .891. And this added to the former product, gives 1.833; or the value of the life for 2 years.—$\frac{473}{499}$ multiplied by .8890, (the value of 1 *l*. due at the end of 3 years) is, .841; and this product, added to 1.833, makes 2.674, or the value of the given life for 3 years.

When the term exceeds 10 years, the rule in Queft. VI. will give thefe values with fufficient exactnefs; and it would do the fame in all cafes, were the values of lives given true to 3 or 4 places of decimals, and in ftrict agreement to the Tables of Obfervation ufed.

The remark now made is to be extended to the values of *joint* lives for given terms. For thefe values, like thofe of *fingle* lives, cannot be found in folving thefe Queftions with fufficient accuracy, when the terms are fmall, by any method, except the tedious one, of multiplying the probability that the 2 lives fhall *both* continue 1, 2, 3, &c. years, by the value of 1 *l*. due at the end of 1, 2, 3, &c. years, and taking the fum of the products in the manner juft defcribed.

Remark II.

If the annuity is to be entered upon, in case of the failure within a given time of any life or lives, *at the end of that time*; and not *at the end of the year in which the failure may happen*; its present value will be the product arising from the continual multiplication by one another of the perpetuity increased by unity; the value of 1 *l*. due at the end of the given time; the annuity; and the probability that the life, or lives, shall fail within the given time. And care should be taken not to confound these two forts of Questions with one another.—Thus; the value in one payment of 10 *l*. *per ann*. to be entered upon eleven years hence, in case a person aged 34 should not live so long, is 26, (the perpetuity increased by unity, interest being at 4 *per cent*.) multiplied by .6496, and by 10 *l*. and also by $\frac{104}{100}$; or 34.8.—This value, divided by 1.04, is, 33.5, the value of an equivalent sum, or of 250 *l*. to be obtained on the same conditions.

The value of the *assurance* of any annuity on the whole continuance of any single life is, by Quest. X. the *excess* of the perpetuity above the value of the life, multiplied by the annuity. And in like manner; the value of the *assurance* of any annuity on the whole continuance of any two *joint* lives, or the *longest* of two lives, is the excess of the *per-petuity*

petuity above the value of the joint lives, or of the longest of two lives, multiplied by the annuity. This is very obvious; but no general method has been yet explained of finding the values of *assurances* on lives and survivorships for terms of years less than the whole continuance of the lives. For this reason, I have been here more explicit than I should otherwise have been; and, as such assurances are now much practised, and may be very useful if their values are rightly determined, I have thought proper to add the two following Questions, which, when joined to Question XI. and Mr. *Simpson's* 33d *Problem* given in the note, p. 39, will, I believe, exhaust this subject as far as two lives can be concerned.

Question XV.

" B, expectant, will lose a given sum,
" should he survive A, *within a given time.*
" What ought he to pay for the *assurance* of
" it?"—In other words: " What ought he
" to pay for a given sum to be received at
" the death of A, should he happen to sur-
" vive him within a given time?"

Answer.

Divide the *sum* of the decrements of life in the Table of Observations from the age of A, for the given time, by the given time; and, by the *quotient*, divide the number of the living in the Table at the age of A; and again, by this *second* quotient (*a*), divide the given sum, reserving the *third* quotient.

Find the value of an annuity on the life of B, for the given time. To this value add the *quotient*, that will arise from dividing the value of an annuity certain, for the given time, by twice the *complement* of the life of B; and the *sum*, multiplied by the *reserved quotient*, will be the required value in a single present (*b*) payment.

EXAMPLE.

Let the Table of Observations be Mr. *Simpson*'s for *London*, or Table VIII. Let the rate of interest be 3 *per cent*. A, seven years of age. B, 30. The given time 14 years. The given sum 100 *l*.—The sum of the *decrements*, in Table VIII. for 14 years from the age of seven, is 73, which, divided by 14, gives 5.2. The number of the living at seven is 430, which, divided by 5.2, and 100 *l*. divided by the quotient, gives *l*. 1.21, the *quotient* to be *reserved*.

(*a*) When the age of A is under 60, and the term so large as to exceed the difference between it and 70, it will be best when the *London* Table is used, to divide the given sum, not by the second quotient here mentioned, but by the *complement* of the life of A, taken out of Table IX.

(*b*) See the demonstration of this rule, and also of the rule that will be given for solving the next Question; in the Appendix, note (H).

The

The value of an annuity for 14 years on the life of B, is, by Queſt. VI. 9.5.—The value of an annuity certain for 14 years, is, by Table II. 11.296, which, divided by 94.4, (twice the *complement* of the life of B, by Table IX (*a*), gives .12, which, added to 9.5, gives 9.62; and this again multiplied by 1.21, the *reſerved quotient*, gives 11.64, the *preſent* value in one payment of 100 *l.* payable at the death of A aged 37, to B aged 30, ſhould A die and leave B the ſurvivor within 14 years.

The preſent value for 14 years of two joint lives, one 7 and the other 30 years of age, may be found, by the help of Table XI, and the rule in the *Scholium* to Queſt. VI. to be nearly 9 years purchaſe; and, *l.* 11.64 divided by this value with unity added, or by 10, gives 1.164, the foregoing value in *annual payments* during the joint lives for 14 years, the *firſt* payment to be made immediately, and the *laſt* payment at the end of 14 years, ſhould the joint lives not fail.

SCHOLIUM.

It deſerves particularly to be remembered, that in this method likewiſe may be calculated, what ſums ought to be paid on any ſurvivorſhip, within a given time, of one life

---

(*a*) This Table gives the *expectations* only, but it ſhould be remembered, that twice the *expectation* is always the *complement* of a life. See note, p. 37.

beyond

beyond another, in confideration of any given fum now advanced.—The following Example of this is a cafe which has offered itfelf in practice.

"A perfon, aged 30, has in expectation an eftate which is to come to him, provided he furvives a *minor*, aged 7, before he is out of his minority; that is, provided he fhould be himfelf living at the time of the minor's death, fhould that happen before he is 21.—In thefe circumftances, he wants to borrow 1000 *l*. on his *expectation*. What *reverfion* out of the eftate depending on fuch a furvivorfhip, is a proper equivalent for this fum now advanced, intereft being reckoned at 3 *per cent*. and the probabilities of life being fuppofed the fame with thofe in Mr. *Simpfon*'s Table of *London* Obfervations?"

ANSWER.

It appears from what has been juft determined, that for *l*. 11.64 now advanced, the proper equivalent in fuch circumftances, is, 100 *l*. to be paid, in cafe the furvivorfhip fhould take place; or, by the *correction* in page 34, as much of the eftate as 100 *l*. will buy at 3 *per cent*. fuppofing the firft rent to be received immediately; (that is, fuppofing the eftate worth 34.33 years purchafe.) or *l*. 2.912 *per annum*.—By the rule of proportion, therefore, for 1000 *l*. the proper equi-

equivalent will be 8591 *l.* in money, or 250 *l.* *per annum* out of the estate.

QUESTION XVI.

" 100 *l.* will be lost to B's heirs, should he
" happen to die after A, *within a given time*.
" What is the price of the *assurance* of it?—
" That is: What is the present value of
" 100 *l.* payable at the death of B, provided
" his death should happen *after* A's death,
" *within a given time?*"

ANSWER.

Divide the sum of the decrements of life in the Table of Observations from the age of B, for the given time, by the given time; and by the *quotient* divide the number of the living at the age of B; and again, by this *second quotient* (*a*), divide the given sum, reserving the *third quotient*.

Find the value of an annuity on the life A for a number of years, less by *one* year than the given time, which subtract from the value of an annuity certain for the same number of years. Multiply the *remainder* by the *reserved quotient*, and divide the *product* by the amount of 1 *l.* for one year; and let this be a *second* reserved quotient.

(*a*) Or rather, if the *London* Table is used, by the *complement* of the life of B, when his age is under 60, and the term exceeds the *difference* between it and 70.

Again.

Again. Multiply into one another the *first* reserved quotient, and the value of an annuity certain for the given time; and divide the product by twice the *complement* of A's life. This *last* quotient, added to the *second* reserved quotient, will be the *answer* in a present single payment.

EXAMPLE.

Let the age of B be 40. Of A 30. The sum 100 *l*. Rate of interest 4 *per cent*. The given time 20 years. The Table of Observations, Mr. *Simpson*'s, or Table VIII.—The sum of the decrements of life, in this Table, from the age of 40 for 20 years, is 127, which, divided by 20, (the given time) gives 6.38.— The number of the living at 40 is 229, which, divided by 6.38, gives 35.8; and 100 *l*. (the given sum) divided by 35.8, gives 2.79, the *first* quotient to be reserved.

The value of an annuity for 19 years on a life at 30 years of age, is 10.3; which, subtracted from 13.134, (the value of an annuity *certain* for 19 years, by Table II) and the remainder multiplied by 2.79, gives 7.89. This product divided by 1.04, (the amount of 1 *l*. in one year) gives 7.60; the *second* reserved quotient.

2.79 multiplied by 13.59, (the value of an annuity certain for 20 years) gives 37.916; and this *product* divided by 94.4, (twice the com-

complement of A's life by Table IX.) gives .401, which, added to 7.60, gives 8 *l.* the *Anfwer*; or, the value of 100 *l.* payable at the death of B, on the contingency of his furviving A aged 30, and *both* dying in 20 years.

It is plain, that this is likewife the fum that ought to be lent to B now, on the expectation of 100 *l.* at his death, provided it fhould happen after A's death in 20 years.

This rule gives the juft folution in all cafes, except when B, the expectant, is the *youngeft* of the two lives, and at the fame time the term of years *greater* than the complement of A's life. In this particular cafe the following rule muft be ufed.

Find, by the preceding rule, the value of the affurance of the given fum for a term of years, equal to the complement of A's life, and let this value be *referved*. Multiply by one another the given *fum*; the *value* of 1 *l.* to be received at the end of a number of years equal to the complement of A's life; and the value of an *annuity certain* for as many years as the given term exceeds this complement. And the *product*, divided by the complement of B's life, and the *quotient* added to the *value referved*, will be the true value fought.

EXAMPLE.

Let the age of B be 30; of A 40. The term 47 years; and every thing elfe as in the

the last Example. The complement of A's life, is, by Table IX, 39.2. The value of 100 *l*. to be received at the death of B, if he survives A within 39 years, may be found by the preceding rule to be *l*. 16.15; the value to be reserved.—The value of 1 *l*. to be received at the end of 39 years is, by Table II, .2166: The value of an annuity *certain* for 8 years; (the excess of the given term above the complement of the life of B by Table IX.) is, 6.733.

And these two values multiplied by one another, and by 100 *l*. give 145.83; which, divided by 47.2, (the complement of the life of B) and 16.15, added to the quotient, make *l*. 19.23, the value sought.

REMARK.

As after finding the present value of an estate, or annuity, it is necessary to *divide* that value by the amount of 1 *l*. in one year, in order to find the present value of a *sum* equivalent to the *annuity*; so, after finding the value of a sum, it is necessary to *multiply* that value by the said amount, in order to find from it the value of an equivalent annuity.

In the first example, therefore, the value of an estate of 4 *l. per annum*, would be *l*. 8.32. In the second Example, 20 *l*. And this is, as it ought to be, the value for the whole duration of the lives, agreeably to the Problem in the note p. 37.

In folving this Queftion, care alfo muft be taken not to forget the *firft* Remark under the foregoing Queftion.

In the fame way with that in which the rules in the three laft Queftions have been difcovered, it is poffible to find rules for calculating the values of *affurances, for given terms*, on lives and furvivorfhips, where three or more lives are concerned. But this is of lefs importance; and I chufe to leave to others the further profecution of this fubject.

CHAP.

CHAP. II.

Containing an Application of the Questions in the foregoing Chapter to the Schemes of the Societies in Great Britain, *for making Assurances on Lives and Survivorships, and for granting Annuities to Widows, and to Persons in old Age.*

SECT. I.

Of the London Annuity, *and the* Laudable *Societies for the Benefit of Widows.*

THE scheme mentioned in Quest. VIII. is nearly that of the *London Annuity* Society. The *Laudable Society* is also formed on a similar plan. In both, the *annual contribution* of every member is five guineas, payable half-yearly; and for this a title is given to an annuity of 20 *l.* to every widow during widowhood, if the husband, after admission, lives *one* year according to the *first* scheme; or *three* years according to the (*a*)

(*a*) In this society a member who lives but *one* year, is entitled to no more than an annuity of 10 *l.* for his widow; if he lives two years, 15 *l.* if he lives three years, 20 *l.* four years 25 *l.* seven years, 30 *l.* ten years, 35 *l.* thirteen years, 40 *l.*

second; of 30*l.* if the husband lives *seven* years, according to both schemes; and 40*l.* according to the *first* scheme, if he lives 15 years, or 13 years, according to the second.—In both schemes also, there is no other premium or fine required, than five guineas extraordinary, at admission, from every member whose age does not exceed 45. The *Laudable Society* admits none above 45, and the *London Annuity* Society obliges every person between 45 and 55 to pay, at admission, five guineas extraordinary, for every year that he is turned of 45.

These are the main particulars in these schemes; and, therefore, both of them, were the annuities to be enjoyed for life, would receive (supposing the members all under 46 at admission, and of the same ages with their wives, and money at 4 *per cent.*) but little more than three-fifths of the true value of the annuities: or about one half, supposing wives, one with another, 10 years younger than their husbands; as appears from Question VIII.

It appears further in that Question, that, supposing the annuities to be *life* annuities, and men and their wives of equal ages, the expectation to which an annual payment of five guineas beginning immediately, entitles, is nearly 14*l.* if the contributor lives a year, and 20*l.* if he lives seven years (*a*), taking

(*a*) The same annual payment will, on the same suppositions, entitle to 14*l.* if a member lives a year, and 18*l.* if he lives *three* years.

F the

the medium between the *London* and the other Tables of Observations.

It is likely, that many persons will be very unwilling to believe, that these schemes are so deficient as they have been now represented. I will, therefore, endeavour to prove this in a way which, tho' less strict, is sufficiently decisive, and may be more likely to be intelligible to persons unskilled in mathematical calculation.—I shall here confine myself to the scheme of the *London Annuity* Society. The differences between it and the scheme of the *Laudable* Society are inconsiderable, and what shall be said of the one will be fully applicable to the other.

According to this scheme, as it has been just described, all that live 15 years in the society will be entitled to annuities of 40 *l. per annum* for their widows. Suppose the whole society, at admission, to be men of 40 years of age, taken one with another. A person of this age has an even chance of *living* 23 years; and he has an even chance of continuing with a wife of the same age, (that is, of continuing in the society) 13 years and ¼ (*a*). Not much less, therefore, than half

(*a*) This is the exact truth according to Mr. *De Moivre*'s hypothesis, and the *Norwich* Table. But according to Dr. *Halley*'s and the *Northampton* Table, a man 40 years of age has an even chance of living no more than 22 years, and of joint continuance with a wife of the same

half the members will continue in the society 15 years; and, consequently, not much less than half the widows that will come upon the society will be annuitants of 40 *l. per annum.* These widows, however, being older than the rest when they commence annuitants, will continue on the society a shorter time; and, therefore, the number constantly in life together, to which they will in a course of years increase, will be proportionably smaller. Putting every thing as favourably as possible, let us suppose, that out of 20 annuitants constantly on the society, *five* will be annuitants of 40 *l. six* of 30 *l.* and *nine* of 20 *l.* To 20 annuitants then the society will pay 560 *l. per annum,* or the 20th part of this sum, that is 28 *l.* to *every* annuitant at an average. But such an annuity for a life at 40, after another equal life, provided both survive one year, is worth (by Quest. VII. p. 24.) in a single present payment, 85 *l.* nearly, according to the *London,* and all the Tables of Observations, interest being all along supposed at 4 *per cent.*

It cannot appear improbable to any one, that this should be the true value of such a reversion. It is not probable, that there is any situation in which the decrements of life

same age, 13 years.—Forty must be more than the mean age of the members of the society at admission, and on this account the number of annuitants of 40 *l.* must be proportionably greater. The mean age, therefore, has been taken very moderately.

are such as can make it a tenth part more or less.—85 *l.* in present payment is the same with 3 *l.* 8 *s. per annum* for ever.—But is an annual payment of five guineas, which must cease as soon as either of two lives each 40 fails, equal in value to such a perpetuity? Every one must see, that there is a great difference.—A set of marriages between persons all 40, will, according to the probabilities of life in Dr. *Halley*'s Table, last, one with another, 15 years (*a*); and an annual payment beginning immediately, during the joint continuance of two persons of this age, is worth 10 years purchase (*b*). The comparison then, in the present case, is between 3 *l.* 8 *s. per annum for ever,* and five guineas *per annum for* 15 *years;* or between an annuity of 3 *l.* 8 *s.* worth 25 years purchase,

---

(*a*) See the beginning of Essay I.

(*b*) The value of such an annual payment, by Table XI, or the *London* Observations, is 9.1; and 10.8, by Mr. *De Moivre*'s hypothesis.—I have not taken into this account the five guineas *fine* paid at admission, because it is obviously of too little consequence to make any considerable difference. The allowances I have made in favour of these schemes are more than equivalent to it. In particular; it should be remembered, that the calculations suppose, that the payments required by these schemes, are yearly payments beginning immediately; (see p. 28) and that, the first payment of the annuity is not to be made 'till the end of the year in which the husband shall die; and also, that the annuity is to be paid yearly, and nothing to be due for any part of the year, in which the annuitant shall happen to die.

and

and an annuity of five guineas worth only 10 years purchase.

But to throw this subject into another light.

Let the number to which the society is kept up be supposed to be 200. It has been demonstrated in Quest. II, that at least half this number of widows will in time come to be constantly on the society; and it has also been just now shewn, that the medium of annuities, payable to them, will be at least 28*l*. After a course of years, then, the society will have a constant expence to bear of 2800*l. per annum.*—But what will be its income?—In order to determine this, we must consider, that there are two sources from whence its income will be derived. First, the annual payments of the members. And, secondly, the money accumulated, or the *capital* raised during the time the number of annuitants is coming to a *maximum*.— The first of these sources affords 1000 guineas, or 1050*l. per annum.* This wants 1750*l.* of the annual expence just mentioned; and, therefore, in order that the income of the society may be equal to the burden upon it, when the annuitants come to a *maximum*, there must be a fund raised in the mean time equal to 43,750*l.* or to an estate in perpetuity of 1750*l. per annum.*—But 1050*l. per annum* beginning immediately, and forborn 25 years, and improved, without loss or delay, all that time at 4 *per cent.*

compound intereſt, will but juſt raiſe ſuch a capital (*a*). There is, therefore, the fulleſt proof, that the ſcheme I am conſidering, is extremely deficient. The truth is, that ſcarcely a *third* of ſuch a capital could be raiſed, as will appear from the following obſervations.

Out of 200 perſons, all 40 years of age, *more* than five, according to the *London* Table of Obſervations, and not ſo many by Dr. *Halley*'s Table, may be expected to die in a year. Suppoſe then five to be the real number of members that will die the firſt year of the ſociety. In ſubſequent years the collective body of members will be continually growing older; and, therefore, the proportion of them that will die every year, will be continually increaſing, 'till it gets to a *maximum*. I will, however, ſuppoſe, that

---

(*a*) Every Queſtion of this kind may be eaſily ſolved in the following manner. In Table I, find the value of 1 *l.* payable at the end of any number of years; and any given annuity divided by that value, will be the annuity to which the given annuity will in that number of years increaſe.—Thus; the preſent value of 1 *l.* payable at the end of 25 years, is .3751, reckoning intereſt at 4 *per cent.* and 1050 *l. per annum* divided by .3751, gives 2,800 *l. per annum*, the increaſed annuity ariſing from 1050 *l. per ann.* In the ſame manner it may be found, that the ſame annuity, forborn 11 years, will increaſe to 1610 *l. per annum.*—But a more particular account of this will be given in the rules annexed to the Tables at the end of this work.

during the first 20 years no more than the number just specified will die every year; and that, consequently, no more than *five widows* will come every year on the society. The ages of all these widows, when they commence widowhood, will, it is evident, be between 40 and 60. One with another then, they may be considered as having commenced widowhood at 50 years of age. Now, five widows left every year at this age, will, in 10 years, increase to 43 constantly in life together, according to the expectations of life in Tables III, IV, and V; and, in 20 years, to 70 (*a*). Suppose the true number alive together at the end of 20 years to be only 62, the greater part of these will be annuitants of 30*l.* and 40*l. per ann.* and the rest 20*l.* Were the former only equal to the latter, the medium of annuities payable to them would exceed 25*l.* Suppose then

---

(*a*) Every calculation of this kind is easily made by the rule in note (A) in the Appendix.—I have put the number living together at the end of 20 years at 62, not only that the reader may be better satisfied that I have kept low enough, but also to make an allowance for such widows as will be left by those members who die within a year after admission, and who, therefore, according to these schemes, will be entitled to no annuities. This allowance is too large: For, after the first year of the scheme, it will not happen above once in 4 or 5 years, that the death of a member will be so circumstanced, supposing the probability that a man at 40 will live a year, to be, as all but the *London* Tables make it, 50 to 1.

this medium to be no more than 26*l*. and it will follow, that, at the end of 20 years, the fociety will have an annual rent to pay of 26*l*. multiplied by 62, or 1612*l*. and, if then able to bear fuch an expence, it muft, in the intermediate time, have acquired an increafe of income equal to the difference between 1050*l*. and 1612*l*. *per ann*. That is; it muft, with its favings, have accumulated a ftock equal to 562*l*. *per ann*. and worth 14,050*l*. But as, during this time, there will be a number of annuitants conftantly increafing, to whom yearly payments muft be made, the favings of the fociety cannot certainly be one half of what they would have been had it been all the time free from all burdens. Suppofe then the ftock produced by thefe favings, to be equal to the ftock that would arife from an income of 1050*l. per ann*. beginning immediately, and improved perfectly at 4 *per cent*. compound intereft, for half the time I have mentioned, or for 10 years, without being fubject to any checks or deductions: fuch an income thus improved, would, in 10 years, produce an additional income of 560*l. per annum*, or a capital of 14,000*l*.—According to thefe Obfervations, therefore, the annual income of the fociety at the end of 20 years, and before a third part of the higheft annuitants could come upon it, would begin to fall fhort of its expences. About that time then

then it would necessarily run aground; and long before the number of annuitants could rise to a 100, it would spend its whole stock, and find itself under a necessity of either doubling the annual payments of its members, or of reducing the annuities one half.

All I have now said is meant on the supposition, that the society begins with 200 members at 40 years of age, and is afterwards limited to that number, by admitting no more new members than will just supply the vacancies occasioned by the loss of old members. If it is allowed to increase, it may continue a longer time. And, for this reason, a society that wants half the income necessary to render it permanent, may very well subsist, and even prosper for 30 or 40 years.—Thus, the *Laudable Society*, was it to keep to its present number of members, might possibly feel no deficiencies for 20 or 30 years to come; but if it should continue to increase at the rate of 70 or 80 every year, it would, at the end of that time, possess a balance so much in its favour, as might enable it to support itself for 20 or 30 years more (*a*). But bankruptcy would

---

(*a*) What has been before demonstrated in Quest. III. should be here recollected, that the number of annuitants on such a society as this, must go on to increase for more than 100 years, after acquiring its greatest number of members.

The *Laudable* Society, I am informed, took its rise from a calculation contained in a pamphlet entitled, *The Pos-*

would come at laſt, and with the more terrible weight the longer it had been deferred.

The rule in the *London Annuity* Society, which obliges every perſon between the ages of 45 and 55, to pay at admiſſion 5 guineas extraordinary, for every year that he exceeds 45, is an advantage to it, but it is a very inadequate, and alſo a very unequitable advantage. For at the ſame time, that it obliges a perſon 55 years of age, to give *more* than the value of his expectation, it takes *above* two-fifths *leſs* than the value from a perſon who is 45 years of age.

If any perſons remain ſtill doubtful about what I have ſaid, I muſt beg their attention to one further argument.

Poſſibility and Probability of a SCHEME *intended for the Benefit of Widows being able to ſupport itſelf.* The ſcheme here referred to, is the ſame with that which this Society has ſince followed; and I am afraid I ſhall not be credited, when I ſay, that the calculation to prove its capacity of ſupporting itſelf, is founded on the ſuppoſition, that a hundred married men, whoſe common age is 36, will leave but *one* widow every year, tho' at the ſame time it is ſuppoſed that two of them will die every year.

This miſtake has made the whole calculation one half wrong.—Nothing can be plainer than that, if the death of a married man does not leave a widow at the end of the year, the reaſon muſt be, that both himſelf and his wife have happened to die in the year. But it is always very improbable this ſhould happen.

(*a*) At 3 *per cent.* the period of doubling money by compound intereſt, is nearly 23 years. At 5 *per cent.* 14 years.

It

It muſt be reckoned upon that every other member of theſe ſocieties, ſuppoſing them to conſiſt of perſons all of the ſame ages with their wives, will leave widows to whom, one with another, (as already ſhewn) at leaſt 28 *l*. *per ann.* muſt be allowed, for as many years as there have been payments from each member. For every 10 guineas then received they muſt ſome time or other hereafter pay 28 *l*. But let it be well conſidered what can enable them to do this. Did money bear no intereſt, for any given ſum now received, they could not afford at any time hereafter to pay more than an *equal* ſum. That is; (ſince the duration of *ſurvivorſhip* is in the preſent caſe, by Queſt. II, equal to the duration of *marriage*) the proper conſideration for any given reverſionary annuity, to be allowed to *all* the ſurvivors of a ſet of marriages, would be, an equal annuity payable by each marriage during its exiſtence; and juſt *half* the reverſionary annuity, if it is to be allowed only to half the ſurvivors, or to widows excluſive of widowers. The annual payment then of *five* guineas, during marriage, can entitle widows to no more than an annuity of *ten* guineas, ſuppoſing money to bear no intereſt. But if money does bear intereſt, the ſame payment will entitle them to more, in proportion to the degree in which it is capable of being improved, during the time between that in which the annual payments begin,

gin, and the commencement of widowhood. Now, it is eafy to fee, that unlefs money bears very high intereft, this improvement cannot be likely in any circumftances to produce a capital, the intereft of which fhall be equal to the annual payment itfelf. Any given annual payment perfectly improved at 4 *per cent.* compound intereft, requires 17 years to double itfelf, fuppofing the firft payment made immediately; or, near 18 years (*a*), if the firft payment is not made 'till the end of a year. But no marriages are *likely* to laft fo long, except thofe among perfons who are very young. A marriage between two perfons, both 40, will not *probably* laft longer than 13 years, according to the probabilities of life in Dr. *Halley*'s Table. A marriage between two perfons, both 50, will not probably, by the fame Table, laft longer than *eleven* years; nor a marriage between two perfons, both 30, longer than 16 years. Such marriages, it is true, may pofiibly laft 30 or 40 years. But this circumftance is more than balanced by the fact, that no lefs poffibly they may not laft *one* year. The annual payments, then, being incapable of fuch an improvement as fhall produce an additional income equal to themfelves; it is obvious, that no fociety ought to go fo far as to

---

(*a*) At 3 *per cent.* the period of doubling money by compound intereft, is nearly 23 years. At 5 *per cent.* 14 years.

allow to widows annuities twice as great as those which might be allowed, supposing no interest of money (*a*); so far, for instance, as to allow, instead of 10 guineas, 20 guineas for an annual payment of five guineas. In the circumstances of most of these societies three-fifths addition may be the full allowance. That is; supposing the annual payment of each member to be five guineas, time may be expected for gaining from hence a capital of 75 guineas, or that shall produce three guineas *per annum* interest; and the proper reversionary annuity will be 16 guineas; or six guineas more than the proper reversionary annuity, did money admit of no improvement.

The preceding observations have gone on the supposition, that the reversionary annuities are to be *for life*. What difference in favour of these societies arises from the circumstance, that the annuities are to be paid only *for widowhood*, cannot be exactly determined. Some judgment, however, may be

(*a*) The money accumulated will not be exactly the same with that to which the annual payment would increase, if improved at compound interest for a number of years, equal to that which the joint lives have an *equal chance* of existing. Much less will the increase be the same with that which would arise from the annual payment forborn, and improved, for a number of years equal to the *expectations* of the joint lives. It will be less than either of these, for a reason explained in note (L) Appendix.

formed

formed of it from what has been said at the conclusion of Queft. II. Were even one half of the widows to marry, still the schemes I have been considering would probably be insufficient. But, in the circumstances of these societies, it cannot be expected, that above one in 10, or perhaps one in 20, will marry. The persons most likely to enter into them, are such as have not the prospect or ability of making competent provisions for their widows in other ways. The widows left, therefore, will in general be unprovided for, and, being also left with families of children, it is quite unreasonable to expect, that any considerable proportion should marry. This is true of such as may happen to be left young; but when a society has subsisted some time, the *greater* part will not be young when left, and these, at the same time that no advantage can be expected from their marrying, will be in general the *highest* annuitants, and, therefore, the *heaviest burdens.*—Moreover, the prospect of the loss of their annuities will have a particular tendency to check marriage among them.—For all these reasons it seems to me likely, that the benefit, which these societies will derive from marriages among their annuitants, will not be very considerable; or at least not *so* considerable as to be equal to the advantages I have allowed them, by calculating on the suppositions, that the money they receive will be *always improved perfectly, without loss or delay, at the rate*

of

of 4 *per cent. compound interest*; that the probabilities of life among males and females are the same, and all husbands likewise of the same ages with their wives, and that consequently the *maximum* of widows on such societies can amount to no more than half the number of marriages (*a*).—With respect to the last of these suppositions, it deserves to be particularly observed, that from accounts taken annually with great care in *Scotland*, it appears, that the widows of the *ministers* and *professors* there (*b*), notwithstanding the diminution occasioned by their marrying, do exceed considerably the number of marriages.

(*a*) Care should be taken in these societies, not to judge of the proportion of widows that will marry, from the proportion that may happen to marry during their first years. For most of the widows that will be left at first will be young; whereas the greater part will not be young when they commence widowhood, after a society has subsisted 30 or 40 years; and, therefore, though one in three or four should marry at first, it will not be reasonable to expect, that half so many should marry after the affairs of the society become stationary.

(*b*) The number of married ministers and professors, for 17 years, from 1750 to 1766, was at a medium 667. And, from 1749 to 1771, the reports have given about 380 as the number of widows all living at one time derived from this whole body. The medium of widows left annually has, for the last 27 years, been $19\frac{1}{4}$; and, for 10 years, ending in the year 1767, but nine of these had married.—Of the annuitants likewise (about 160 in number) on the fund established among the Dissenters in *London*, for relieving the widows of indigent ministers, it is found that few ever marry. See Chap. 2. Sect. 2. See likewise the latter end of the 4th Essay; and note (A) in the Appendix.

And

And certainly it would be unreasonable in these societies not to reckon that the same will happen among them.—Indeed it seems certain, that notwithstanding the hazards that attend child-bearing, the probability, that the woman shall survive in marriage, and not the man, is much greater (*a*) than is commonly imagined. It will be shewn in the last Essay, that it is not less than the odds of 3 to 2; and had I calculated agreeably to this fact, the values of annuities for widows, would have been given near a quarter greater than they have been given on the supposition, that the chance of survivorship is equal between men and their wives.—It must be added, that I have made no account of any expences attending the execution and management of the schemes of these societies. Some such expences there must be, and some advantages should be always provided in order to compensate them.

There are in this kingdom several institutions for the benefit of widows, besides the

---

(*a*) Partly, as observed in page 8, on account of the greater mortality of males, but chiefly on account of the excess of age on the man's side.—According to the printed articles of agreement, the *Laudable Society* pays no regard to this excess of age; and the allowance required on this account by the *London Annuity* Society is so trifling that it deserves no notice.

In March 1770, thirty-two husbands had died in the *Laudable Society*, and 27 wives. They seem, therefore, to be already beginning to experience, that the chances of survivorship in marriage are in favour of the wife.

two

two on which I have now remarked; and in general, as far as I have had any information concerning them, they are founded on plans equally inadequate. The motives which influence the contrivers of these institutions are, without doubt, *laudable*; but they ought, I think, to have informed themselves better. This appears sufficiently from what has been said; but I will just mention one further proof of it.

The *London Annuity Society* promises that, if in 21 years; and the *Laudable Society* that, if in 25 years, it shall appear that there has been all along an annual surplus in favour of the societies, it shall be employed in either raising the *annuities*, or in sinking the *annual payments*. Now, they may be assured, that, if at the end of these periods, they should not be possessed of a considerable surplus, the true reason will be, their having granted much higher annuities than the annual contributions are able permanently to support: For it has been demonstrated, that the number of annuitants, and consequently the amount of the annual expences, will go on increasing for a long course of years beyond these periods. The effect, therefore, of carrying into execution this regulation will be, precipitating that bankruptcy which would have come too soon had there been no such regulation.

It has been said in defence of these Societies, that the deficiencies in their plans cannot be of much consequence, because their rules oblige them to preserve a constant equality between their income and expences, by reducing the annuities as there shall be occasion. And from hence it is inferred, that they can never be in any danger of a bankruptcy.—In answer to this, it has appeared, that the time when they will begin to feel deficiencies is so distant, that it will be too late to remedy past errors, without sinking the annuities so much, as to render them inconsiderable and trifling. All that is given too much to *present* annuitants is so much taken away from *future* annuitants. And if a scheme is *very* deficient, the first annuitants may, for 30 or 40 years, receive so much more than they ought to receive, as to leave little or nothing for any who come after them. Deficient schemes, therefore, are attended with particular injustice; and this injustice will be the same, if, instead of *reducing* the annuities, the annual payments should be increased; for all the difference this can make will be, to cause the injustice to fall on *future contributors*, instead of *future annuitants*.

But what requires most to be considered here is, that, after either the annuities have been for some time in a state of reduction, or the contributions in a state of increase, it will be seen that these Societies have gone upon wrong plans, and, therefore, they will be deserted and avoided; the consequence of which will prove still greater deficiencies in their

their annual income, and a more rapid desertion and decline, 'till a total dissolution and bankruptcy take place.—This will be the death of most of the present Societies for providing for widows, if they continue to be encouraged, and do not soon alter their plans: And at that period the number of *annuitants* will be greater than ever; whose annuities, having no other support than the poor remains of a stock always insufficient, will be soon left, without the possibility of relief, to lament that ignorance and credulity which gave rise to these societies, and which had so long supported them.

In the *London* Annuity Society, there is an encouragement to *batchelors* and *widowers* to join them, arising from the additional annuities to which they will be *immediately* entitled, when they marry, in consequence of having made their payments a greater number of years; and it is imagined, that particular advantages will be derived from such members. But even these will in general pay much less than the value of their expectations.—A person who begins an annual contribution of five guineas at the age of 24, will, should he live 11 years, and marry a woman of the same age at the end of that time, entitle her immediately to 35 *l.* *per ann.* during survivorship, and to 41 *l. per annum* should he live four years after marrying,

ing, (interest being at 4 *per cent.*) *(a)*. In this particular case, therefore, a person will pay nearly the true value of his expectation. But *all* at all ages who *marry*, and *most* of those who *die*, in less time than 11 years after admission, will pay less than the value of their expectations.

SECT. II.

Of the Association among the London *Clergy, and the Ministers in* Scotland, *for providing Annuities for their widows.*

IN April, 1765, the clergy within the bills of mortality, and the county of *Middlesex*, at a general meeting in *Sion-College*, agreed to form themselves into a society for the support of their widows and orphans. Many in this respectable body may be capable of doing, in a better manner, what I have attempted in this Treatise; and they are, perhaps, already sensible of the deficiencies in the plan

---

(a) The value of five guineas *per annum* (first payment made immediately) for 11 years, subject to failure should a life now 24 fail; and, after 11 years, for the joint lives of two persons both 35, is, by the Table of *London Observations*, *l*.69.3 — By Dr. *Halley*'s Table, *l*. 76.44. — The present value of 35 *l. per annum* for life to the widow of a person now 24, should he live 11 years, and marry a woman of the same age with himself at the end of that time; and also of 6 *l*. more, or 41 *l. per annum* in all, should he live after marriage four years; is, by the Table of *London* Observations, *l*.69.36. — By Dr. *Halley*'s Table, *l*. 76.03.

which

which they have established. I shall not, however, I hope, do wrong, in taking the liberty to recite briefly this plan, in order to introduce a few observations upon it.

According to the printed articles, every clergyman possessed of any benefice, lecture-ship, or licensed curacy, within the bills of mortality, and the county of *Middlesex*, who subscribes annually one guinea, or two guineas, or more, shall entitle his widow to an annuity; or, if he leaves no widow, he shall entitle any such children as he shall leave, to the same annuity for seven years as his widow would have had. And, in case a widow possessed of an annuity, should either *die* or *marry* before the lapse of 10 years, from the commencement of her annuity, such children of her former husband, as shall be then alive, are to be entitled to as many of the ten years payments of the annuities as she shall not have received.—The annuity is fixed to no particular sum, but instead of this, it is ordered, that a fourth part of the annual subscriptions and interest shall be divided the first three years after the establishment of the society; half only the next four years; and 3-4ths the next five years; provided, however, that in no one of these 12 years the dividend shall exceed 20 *l*. to the widows and orphans of the clergy subscribing two guineas or more; and 10 *l*. to the widows and orphans of the subscribers of one guinea. And, after the expiration of 12 years, the whole amount of the subscriptions, and of the interest of the capital stock, is to be divided proportionably for ever.—It is further provided, that every clergyman, who shall be married, or have children, at the time of his subscription, shall pay a fine of two guineas towards a capital stock, if a subscriber of two guineas or more, and 40 years of age or upwards. If 50 years of age or upwards, he shall pay a fine of three guineas; if 60 or upwards, five guineas. But, if not married at the time of his subscribing, and shall afterwards marry, he shall pay a fine according to the age he shall be of at the time of his marrying. The obligation laid upon all, whether married or unmarried, to become subscribers, is, an incapacity of being admitted members without the consent of a general court, unless, within two years after becoming possessed of any ecclesiastical employment, they subscribe.

Every one who has attended to the observations in this and the preceding chapter, must know what judgment to form of these regulations.

Let us suppose that all the clergy in *London* and *Middlesex* came into this association from the first; and that one with another they are subscribers of two guineas annually; and that there are among them as many unmarried persons as married.

In

In this case, it may be learnt from Queſt. XIII, that the annuity to which widows ſhould be entitled, (ſuppoſing no allowance to the children of any that die) ought not to exceed 10 or 11 guineas at moſt, and that, beſides the annual ſubſcriptions, there ought to have been a fine paid at the commencement of the ſcheme, by every married perſon, of ſix guineas at leaſt, or, by the whole number of ſubſcribers, three guineas. If the number of married members is double the unmarried, the annuity ought not to exceed eight guineas; and the fine from every member ſhould be about four guineas.—The order, that only a fourth part of the annual ſubſcriptions and intereſt ſhall be divided the firſt three years, half the next four years, and three quarters the next five, is without reaſon; becauſe the number of claimants, for the firſt 12 years of the ſcheme, will be ſo few, that it will not be poſſible, during that time, that there ſhould be occaſion for dividing any proportions ſo large of the annual ſubſcriptions and intereſt, unleſs they are indeed beyond all bounds too little. —After 12 years, the number of annuitants will go on increaſing for near 50 years, as appears from Queſt. III. The conſequence, therefore, of dividing, after that time, the whole amount of the annual ſubſcriptions and intereſt, will be a conſtant yearly diminution in the dividends for near 50 years; and making the payments to the first claimants much more considerable than they ought to be, at the expence of all subsequent claimants.—For these reasons; it appears to me out of all doubt, that this scheme is by no means likely to answer the good ends proposed by it; and that, therefore, it will be best to lay it aside. At the time it was settled it was, I find, further agreed, that the annual subscriptions of the *laity*, together with the interest of their benefactions, unless otherwise directed by the donors; and the annual subscriptions of such of the clergy as shall so direct, shall make a *charitable fund* to be applied to the relief of the distressed widows or children of all the clergy within the limits I have mentioned, whether subscribers or not, provided that in no one year of the first twelve more than 20*l.* be given out of the fund to any one family.— This is an excellent design; and if the money arising from all the subscriptions is thrown into this fund, an important means of relief may be provided, for such of the more indigent widows and families as will accept the help of charity.

There is one more scheme of particular consequence, which I must take notice of: I mean, that which is established by Act of Parliament, among the ministers and professors in *Scotland*, for making provision for their widows and orphans. The last mentioned

tioned fcheme, and alfo feveral others of the fame kind (a) in this kingdom, have been formed on the model of this: and the fuccefs with which it has been hitherto attended, is one of the principal caufes to which they have owed their rife. It is, therefore, proper I fhould give fome account of it; and it will be fufficient with this view to mention, "that for an annual payment, which
"begun immediately, of *five guineas* from
"1011 contributors, 667 of whom are mar-
"ried perfons, befides a tax on weddings,
"producing about 142 *l. per ann.* it entitles
"every widow to an annuity of 20 *l.* during
"widowhood, and alfo every family of chil-
"dren that fhall be left by fuch members
"as die without leaving widows, to 200 *l.*"
This fcheme contains a variety of other particulars; but this is its fubftance—It commenced on the 25th of *March*, 1744; and from that time, to the 22d of *November*, 1770, or in 26 years and near 8 months, 151 minifters and profeffors died, and left families of children without widows; that is, 5.66 fuch families were left annually;

(a) There is one among the Diffenting Minifters in the counties of *Chefter* and *Lancafter*, and another among the Diffenting Minifters in *Cumberland, Northumberland, Weftmoreland,* and *Durham.*—Even the *London Annuity* Society, tho' its plan is totally different, profeffes to form itfelf on the principles of the *Scotch* eftablifhment, and to derive encouragement from it.

and

and the annual disbursements to them have therefore been 1132 *l.* Subtract this sum from 5450 *l*, the whole annual income; and the remainder, or 4318 *l. per ann.* will be the standing provision for bearing the expence of all the annuitants possible to be derived from 667 marriages. Such an annual payment, or 4.27 each from 1011 contributors, is the same with 6.55 each, from 667 contributors; and, consequently, it appears, that in this establishment a contribution is received equivalent to an annual payment beginning immediately, of *l.* 6.55 from every married man, in order to entitle his widow to an annuity of 20 *l.* during her widowhood.

In the Societies mentioned in the last section, annuities increasing from 20 *l.* to 40 *l.* are promised to widows for an annual payment of only 5 guineas (*a*). And, in all the societies for the benefit of widows with which I am acquainted, there is an equal or a greater disproportion between the contributions received, and the annuities promised.— With what strange rashness then has the plan of this establishment been copied? And how absurdly have the societies in this kingdom pleaded it as a precedent which encourages and favours them?—It would be trifling to say more on this subject.

(*a*) See page 67.

It may be observed that the annual income for the support of this establishment, supposing it to have only the benefit of widows in view, ought be *l.* 7. 19 *per ann.* from every marriage, according to Quest. XIII. p. 44. and *l.* 7.44 *per ann.* according to the calculation in Note F, Appendix.

These determinations exceed the income actually provided. But the excesses are by no means considerable enough, to afford any certain reason for concluding, that the fund of this establishment will prove insufficient. I was, however, once led to entertain some doubts on this subject. And in these doubts I thought myself confirmed by observing, that, in the calculations (*a*) made at the commencement of the scheme, the number 333 was stated, as the *maximum* of widows living at one time, likely to come upon it, or to be derived from 20 (*b*) widows left annually; and also, that 40 years was stated as the number of years necessary to bring on this *maximum*; whereas I was satisfied, that

(*a*) See Table III. in a book printed at *Edinburgh* in 1748, entitled, Calculations, with the principles and data, on which they are instituted, relative to a late act of parliament, entitled, *An Act for raising and establishing a Fund, for a provision for the widows and children of the ministers of the church, and of the heads, principals, and masters of the Universities of* SCOTLAND; shewing the rise and progress of the Fund.

(*b*) See the beginning of note A, Appendix.—See likewise the note in p. 79.

the *maximum* of widows would not prove much less than 400; nor the number of years neceſſary to bring it on, leſs than 60.—— In the former editions of this work, I gave a diſtinct account of this. But I have lately received ſuch information (*c*) as has convinced me that my doubts have been in a great meaſure groundleſs. I have learnt, in particular, that there have been calculations ſubſequent to thoſe I had ſeen; and that this eſtabliſhment has enjoyed advantages and proviſions for its ſupport which I was unacquainted with, and which give reaſon for expecting that it will indeed be able to bear the expence of 400 annuitants, ſhould ſo many come upon it. I ſhould only tire moſt of my readers, were I to enter into an account of theſe advantages and proviſions. It will be of more importance to take this opportunity to obſerve, that the probabilities of life from which the determinations I have mentioned are derived, though much lower than the probabilities of life among the miniſters and their wives in SCOTLAND (*d*), are yet ſuch as give the values of reverſions depending on ſurvivorſhips among them too high.

(*c*) I owe this information to the kind and very obliging candour of the reverend and ingenious Dr. WEBSTER, of *Edinburgh*.

(*d*) More particular notice will be taken of this at the concluſion of the laſt Eſſay.

In

In order to understand this, it must be considered, that the difference between the probabilities of life in different situations, takes place chiefly in the first and the middle stages of life; and that in old age, they are nearly the same in all situations. This may be deduced with abundant evidence from the three first Tables in the *Supplement* compared with the two last, and with the Table of Observations for LONDON. The effect of this must be to *increase* the duration of *joint lives*, and at the same time to *lessen* the duration of *survivorship* in those situations which are most favourable to health. Or, in other words, to render the duration of marriage in such situations, greater than it would otherwise be in proportion to the duration of widowhood; and, consequently, to reduce the present value in annual payments during marriage, of any given annuity payable during widowhood. For instance. Were the probabilities of life among the ministers and their wives in SCOTLAND the same that they are in Mr. *De Moivre*'s hypothesis, or in Tables III. IV. and V. in the Appendix, the duration of marriages among them, taken one with another, could not be more than 19 years. The duration of widowhood would be 22 years, and the *maximum* of widows living at one time derived from 667 marriages constantly kept up, would be considerably more than 400.—Were the probabilities of life among them the same that they are in LONDON, the duration of marriage would be still *less*, and the duration of widowhood *greater*, and the *maximum* of widows derived from 667 marriages, could not be less than 500. But the fact is, that the duration of marriage among them is 21 years and a half (*a*); and that of widowhood about 20 years. And it appears also, from accounts taken annually, that the number of widows living at one time, derived from the whole body of ministers and professors, is about 380. It is, therefore, certain that a smaller income must be sufficient for the support of this scheme than would be necessary, according to the probabilities of life in the Tables just mentioned.—And upon the whole; after a careful review of all the circumstances of this establishment in its present state, I am well satisfied, that the success with which it has been hitherto attended, is likely to continue; and that it will indeed prove a permanent foundation of that assistance to the *widow* and *fatherless* which is intended by it.—Caution, however, and vigilance, will for some time be necessary. Many more years must pass before it can re-

---

(*a*) See a note at the conclusion of the last Essay; and also note F, Appendix.—The *maximum* of widows (or 380) divided by the number left annually (or 19.2) gives 20, the expectation of widowhood. See p. 79, and note A, *Appendix*.

ceive a decisive confirmation from experience. Events have hitherto favoured it more than could have been reasonably expected. They may perhaps hereafter try it; and deviations from probability may arise, which cannot be now foreseen.—But I ought to ask pardon for making these remarks. The venerable ministers and professors concerned will, I hope, excuse me. They are eminently distinguished by their abilities and knowledge; and can have little need of any information which I am able to give them.

SECT. III.

Of the best Schemes for providing Annuities for widows.

INstitutions for providing widows with annuities would, without doubt, be extremely useful, could such be contrived as would be *durable*, and at the same time *easy* and *encouraging*. The natures of things do not admit of this in the degree that is commonly imagined. The calculations and rules, in the preceding chapter, will enable any one to determine in all cases to what reversionary annuities any given payments entitle, according to any given valuation of lives, or rate of interest. From Quest. VII. and VIII. in particular, it may be inferred that (interest being

being at 4 *per cent.* and the probabilities of life as in Mr. *De Moivre's* hypothesis, or the *Breslaw, Norwich,* and *Northampton* Tables) for an annual payment beginning immediately of *four* guineas during marriage; and also for a guinea and half in hand, on account of each year that the age of the husband exceeds the age of the wife, every married man, under 40, might be entitled to an annuity for his widow, during *life*, of 5 *l.* if he lives a year, 10 *l.* if he lives *three* years, and 20 *l.* if he lives *seven* years. Money can scarcely now in this kingdom be improved at so high a rate as 4 *per cent*. But, perhaps, it might be reasonably expected, that an advantage, sufficient to compensate this disadvantage, would be derived, from changing the annuities I have mentioned into annuities during widowhood. One may, at least, venture to pronounce, that nothing much worse could befall a society that went on this plan, than the necessity of some time or other adding half a guinea to the annual payments.

If such a society chuses, that those who shall happen to continue members the longest time, shall be entitled to still greater annuities, six guineas, additional to all the other payments at admission, would be the full payment for an annuity of 25 *l.* and 12 guineas for an annuity of 30 *l.* if a member should live 15 years.

All

All batchelors and widowers might be encouraged to join such a society, by admitting them on the following terms.—*Four guineas* to be paid on admission, and *three guineas* every year afterwards, during celibacy; and, on marriage, the same payments with those made by persons admitted after marriage; in consideration of which, 1 *l. per annum*, for every single payment before marriage, might be added to the annuities, to which such members would have been otherwise entitled.

For example. If they have been members four years, or made five payments before marriage, instead of being entitled to life-annuities for their widows of only 5 *l.* 10 *l.* 20 *l.* 25 *l.* and 30 *l.* on the conditions I have specified, they might be entitled to annuities of 10 *l.* 15 *l.* 25 *l.* 30 *l.* and 35 *l.* Or, if they have been members nine years, and made 10 payments, they might, instead of the same annuities, be entitled to annuities of 15 *l.* 20 *l.* 30 *l.* 35 *l.* and 40 *l.*—In this case, the contributions of such members as should happen to desert, or die in celibacy, would be so much profit to the society, tending to give it more strength and security.

This is one of the best schemes that I am able to think of, or would chuse to recommend. There are, however, others no less safe and encouraging which some may prefer,

fer, and which therefore, I will juſt propoſe.

Let the probabilities of life be the ſame with thoſe in the Tables juſt mentioned. Let money be ſuppoſed to be improved at no higher intereſt than 3 *per cent*. Let the reverſionary annuities promiſed to widows be 10 *l. for life*, if a member lives five years after admiſſion, and 15 *l*. more, or 25 *l*. in all, if he lives 11 years. The proper payments for ſuch an expectation, from a married man not exceeding 50 years of age, will, in the neareſt and moſt convenient round ſums, be four guineas *in annual payments* beginning immediately, and two guineas in hand for every year that his age exceeds his wife's, not admitting any greater exceſs than 15 years: Or, if the whole value is given in *one preſent payment*, 40 *l*. added to a guinea, for every year that his age falls ſhort of 50, beſides the payment juſt mentioned on account of diſparity of age.—For example. Four guineas in annual payments, beſides 10 or 20 guineas in hand, according as the age of the huſband exceeds the wife's 5 or 10 years. Or, if the whole value of the expectation is given in one payment, 10 guineas added to 40 *l*. (that is 50 *l*. 10 *s*.) from a man whoſe age is 40; and, in like manner, 20 guineas added to 40 *l*. (that is 61 *l*.) from a man whoſe age is 30;

beſides

besides the payment just mentioned on account of disparity of age.

If money is improved at 4 *per cent.* or, on account of any advantages attending a scheme, may be justly considered as so improved, the full payments for the expectation I have mentioned will be about one eighth (or half a guinea) less in the *annual payments* during marriage; and a quarter less in all the other payments. That is: A married man, *at* or *under* 50, would, besides three guineas and half in annual payments during marriage, be bound to add a guinea and half for every year he is older than his wife: Or, if he chuses to give the value of his expectation in one payment; besides the common contribution of 30 *l.* and a guinea and half for every year his age exceeds his wife's; he would be bound to pay three quarters of a guinea, for every year he is less than 50 years of age; that is, 53 *l.* 12 *s.* 6 *d.* in all, supposing him 40 years of age, and 10 years older than his wife.—All these payments doubled would entitle to double annuities.

There is one particular advantage which societies formed on a plan of this kind would enjoy (*a*).—Persons who know themselves subject to disorders, which are likely to render them short-lived, will have no great temptations to endeavour to gain admission into

(*a*) See another advantage mentioned under Quest. VIII, p. 28.

such societies; and, if admitted, the danger from them will be less than on any other plan. Were it not for this danger, one might recommend the following plan, as one of the most inviting.

In the plans hitherto mentioned it is implied, that, if either a member or his wife dies within any of the periods specified, the additional annuities, that would otherwise have become due, will be lost. But it would be much more agreeable to a purchaser, that they should be made certain to his wife, provided she lives to the end of these periods, though in the mean time his own life should fail. The value of such annuities may be computed by the rule in Quest. IX.

Suppose, for enstance, the *scheme* to be " that a wife shall be intitled certainly to a " life-annuity of 20*l*. the first payment of " which shall be made at the end of 12 years, " provided she should be then alive, and her " husband dead; or at the end of any year " beyond this term in which she may hap- " pen to be left a widow." Suppose it also stipulated, " that she shall be entitled to " 10*l*. more, or 30*l*. in all, on the same " terms, provided she should live 16 years." —The value of such an expectation (interest being at 3 *per cent*. and the probabilities of life as in Mr. *De Moivre*'s hypothesis) will be, in the most convenient round sums, supposing none admitted above 50 years of age, seven guineas in annual payments to be continued during marriage, and to begin immediately; besides four guineas in present money for every year, as far as 15 years, that the husband's age exceeds the wife's, if he is between 40 and 50, and three guineas on the same account if he is under 40: or, if the whole value of the expectation is given in one present payment, 70*l.* added to a guinea and half, for every year that the husband's age falls short of 50, besides the payment just mentioned on account of disparity of age.

If the annuities are made to be annuities during *widowhood*, and not during *life*, and the advantage arising from hence, is supposed equivalent to the difference between the improvement of money at 4 *per cent.* and its real improvement; the value of the expectation just mentioned, (that is, its value at 4 *per cent.*) will be six guineas in annual payments; besides three guineas in present money, for every year that the husband's age exceeds the wife's, if he is between 40 and 50; and 2 guineas, if he is under 40: or, if the whole value of the expectation is given in one present payment, 56*l.* added to 1*l.* 5*s.* for every year that his age falls short of 50, besides the payment last mentioned on account of inequality of age (*a*).

He

(*a*) Supposing 16 years the only term, the annuity 20*l.* and interest at 4 *per cent.* the proper payments will be nearly, in the case of equal ages and *single* payments, 46*l.*

He that will give himself the trouble to calculate, agreeably to the directions in the Queſtions to which I have referred, will find that, taking all particular cafes together, the rules now given come as near the truth as there is reaſon to defire in an affair of this nature, the *defects* in some cafes being nearly compenfated by the *exceſſes* in others.

I have calculated here, as well as in moſt other places, from Mr. *De Moivre*'s hypotheſis, becauſe its conformity to the three Tables which I have ſo often mentioned, convinces me, that it gives a proper *medium* between the different values of *town* and *country* lives. In the country the probabilities of life are much higher; but in *London*, and probably in all *great* towns and ſome *ſmaller* ones, they are much lower.

46 *l.*—40 *l.*—29 *l.* as the age of the man is 30, 40, or 50. Or, in *annual* payments, *l.*3.80.—*l.*3.66.—*l.*3.13—Suppoſing the woman's age 10 years leſs than the man's, the fame values will be, in *ſingle* payments, *l.*58.92.—*l.*56.56. —*l.*53.66.—In *annual* payments *l.*4.63.—*l.*5.—*l.*5.41.— It appears, therefore, that a ſociety, ſuppoſing money improved at the rate of 4 *per cent.* might entitle all married men *indiſcriminately*, who are under 50 years of age, to ſuch an expectation as this for their wives, for either 60 *l.* in *one* payment, or five guineas in *annual* payments. —But equity requires, that different payments ſhould be made, according to the different comparative ages of men and their wives; and Tables might be formed for ſhewing, at one view, what theſe different payments ought to be in all caſes. If ſuch Tables are wanting, recourſe muſt be had to ſome ſuch eaſy rules as thoſe I have ſtated above.

It

It is proper to add, that, according to the values of lives and survivorships deduced both from the *London* and Dr. *Halley*'s Table, and taking interest as low as 3 *per cent*. all women whose husbands are under 50 years of age, might be intitled to an annuity of 24 *l*. during *life* (the first payment to be made at the end of the year in which they shall be left widows) for the sum of 100 *l*. supposing 3 *l*. additional given on account of every year that they are younger than their husbands.— At 4 *per cent*. an annuity of 30 *l*. might be granted on the same terms.

In the year 1690, the company of *Mercers*, in *London*, adopted such a scheme as that last mentioned. For 100 *l*. in *one present payment*, they entitled every subscriber to a *life-annuity* for his widow of 30 *l*.; and this, at that time, (when money bore 8 *per cent*. interest) was considerably less than the value of the money advanced, supposing men and their wives of equal ages. As the interest of money sunk, they sunk also the annuity, first to 25 *l*. and then to 20 *l*. and 15 *l*. But at last, after carrying on the scheme for above 50 years, finding the burden of the annuitants too heavy, and likely to go on increasing, they were obliged to drop the scheme and to stop payment. In a little time, however, by a parliamentary aid of 3000 *l*. *per ann*. which they are now enjoying, they were restored to a capacity of making good all their engagements, and of paying their arrears.—Their failure, is, indeed, much to be lamented; for, in consequence of it, the public has lost the benefit of an institution, that for many years promised the happiest effects, by encouraging marriage, and affording relief to indigence. The rapid fall of the interest of money; their admitting purchasers at too advanced ages; and, particularly, their paying no regard to the difference of age between husbands and their wives, must have contributed much to hurt them. Some of the principal causes, therefore, which have rendered them unsuccessful, may be now avoided; and for this reason I should be glad to see some similar scheme, providing, as this did, annuities for *life*, and not for *widowhood*, undertaken. If well planned, it would, I think, be a proper object of *parliamentary* encouragement.

It must, however, be remembered, that the issue of the best schemes of this kind must be in some degree uncertain. For want of proper observations, it is not possible to determine what allowances ought to be made, on account of the higher probabilities of life among females than males. No prudence can prevent all losses in the improvement of money; nor can any care guard against the inconveniencies to such schemes, which must arise from those persons being most ready to fly to them who, by reason of concealed disorders,

orders, feel themselves most likely to want the benefit of them.

The societies, therefore, on which I have remarked in the first section of this chapter, would have reason to take warning from what has happened to the *Mercers Company*, were the schemes on which they are formed perfectly unexceptionable. But I have demonstrated that these schemes are very defective; and that the longer they are carried on, the more mischief they must produce. 'Tis vain (as appears from Quest. III.) to form such establishments with the expectation of seeing their fate determined soon by experience. If not more extravagant than any ignorance can well make them, they *will* go on prosperously for 20 or 30 years; and, if at all tolerable, they *may* support themselves for 50 or 60 years; and at last end in distress and ruin. No experiments, therefore, of this sort should be tried hastily. An unsuccessful experiment must be productive of very pernicious effects. All inadequate schemes lay the foundation of *present* relief on *future* calamity, and afford assistance to a *few* by disappointing and oppressing *multitudes*.

As the persons who conduct these schemes can mean nothing but the advantage of the public, they ought to listen to these observations. At present their plans are capable of being reformed; but they cannot continue so always; for the greater number of exorbitant

bitant payments they now make to annuitants, the more they confume the property of future annuitants, and the lefs practicable a retreat is rendered to a rational and equitable and permanent plan (*a*). They fhould, therefore, *immediately* (*b*) either *reduce* their fchemes, or change them into one of thofe which I have propofed. But, I am afraid, this is not to be expected. The neglect with which they have received fome remonftrances that have been already made to them, gives reafon to fear, that what has been now faid will be in vain; and that thofe who are to come after them, muft be left to *rue* the confequences of their miftakes.

SECT. IV.

Of Schemes for providing Annuities for Old Age.

A General difpofition has lately fhewn itfelf, to encourage fchemes for granting *annuities* to perfons in the latter ftages of life; and this has occafioned the 6th Queftion in the former Chapter; and, as a further and more particular direction in cafes of this kind, I have thought it neceffary here to give the following Table.

(*a*) See p. 82, 83. Sect. I.
(*b*) Thus; was the *London* Annuity Society to make their loweft annuity 10*l*. the next 20*l*. and the higheft 30*l*. they would probably be fafe. But, after proceeding on their prefent plan fome years longer, fuch a reduction would by no means be fufficient. See a farther account of thefe Societies in the SUPPLEMENT.

Annuities for Old Age.

Values of 1l. per ann. for life, after 50, to persons whose ages are	Values in one present payment, interest 4 per cent.	Interest 3 per cent.	Values in annual payments, 'till 50, to begin at the end of a year, interest 4 per ct.	Interest 3 per cent.
10	1.235	2.015	.0789	.113
15	1.583	2.444	.106	.146
20	2.028	2.989	.146	.193
25	2.594	3.644	.203	.259
30	3.369	4.508	.297	.366
35	4.446	5.667	.466	.559
40	5.953	7.232	.822	.950
Values of the same annuity, after 55, to ages			Values in annual payments till 55.	
30	2.114	2.937	.167	.211
35	2.722	3.632	.241	.297
40	3.732	4.708	.394	.464
45	5.088	6.115	.703	.803
Values of the same annuity, after 60, to ages			Values in annual payments till 60.	
35	1.667	2.290	.135	.168
40	2.234	2.923	.203	.245
45	3.043	3.811	.327	.384
50	4.255	5.061	.600	.679

The numbers in the 2d and 3d columns of this Table, multiplied by any annuity, will give the value of that annuity in a *single* payment, to be enjoyed for life, by the ages corresponding to those numbers in the first column, *after* the age mentioned at the head of of that column.—And in the same manner; the numbers in the 4th and 5th columns will give the values in *annual* payments.—Thus; The value of 44 *l. per annum*, to be enjoyed for life, after 50, by a person now 40, (interest at 4 *per cent*.) is 5.95, multiplied by 44, or *l*. 261.9, in a *single* payment; and .822, multiplied by 44, or *l*. 36.16, in *annual* payments 'till 50, the first payment to be made at the end of a year.

In order to find the same values, partly in *annual payments*, and partly in any given *entrance* or *admission-money*; say; " As the va-
" lue of the *given annuity* in a *single* payment,
" (found in the way just mentioned) is to the
" given *entrance-money*; so is its value in *an-*
" *nual* payments, to a fourth proportional;
" which, subtracted from the value in *annual*
" *payments*, the *remainder* will be the annual
" payment due, over and above the given
" entrance-money."

Example.

Suppose a person now 40, to be willing to pay 200 *l*. entrance-money, *besides* such an annual payment for 10 years as shall, together with his entrance-money, be sufficient to entitle him to a life-annuity of 44 *l*. after 50. What ought the annual payment to be?

An-

Answer.

L.8.55.—For, *l.* 261.9, is to 200 *l.* as *l.* 36.16, to *l.* 27.61; which, subtracted from *l.* 36.16, the remainder is *l.* 8.55.

This Table has been calculated from the *probabilities* and *values* of lives in Tables III. and VI. The probabilities of life among the inhabitants of *London*, are (as I have often had occasion to observe) much lower than among the generality of mankind; and the values in the preceding Table, had they been given agreeably to the *London* Observations, would have been less. But, certainly, an office or society, that means to be a permanent advantage to the public, ought always to take higher rather than lower values, for the sake of rendering itself more secure, and gaining some *profits* to balance *losses* and *expences*.

There have lately been established, in *London*, several societies for granting such annuities as those now mentioned; and he that will compare their true values, as they may be learnt from the preceding Table, with the *terms* of admission into these societies, as given in their printed *Abstracts* and *Tables*, must be surprised and shocked. They are all impositions on the public, proceeding from

from ignorance, and encouraged by credulity and folly.

It has been shewn; that the proper payment, (allowing compound interest at 4 *per cent.*) for an annuity of 44 *l.* to be enjoyed by a person now 40, for what may happen to remain of his life after 50, is 200 *l.* in *admission-money*; besides *l.* 8. 5 5, or 8 *l.* 11 *s.* in annual payments 'till he attains to 50, the first of these payments to be made at the end of a year.—The conditions of obtaining this annuity, according to the Tables of the *Laudable Society of Annuitants for the benefit of age*, are 76 *l.* 17 *s.* in *admission-money*; and 6 *l.* 14 *s.* in *annual payments.*—According to the Tables of the society of *London Annuitants for the benefit of age*, the conditions of obtaining the same annuity are 30 *l.* in *admission-money*, and 10 *l.* in *annual payments.*—The *Equitable* Society of Annuitants requires for the same annuity 38 *l.* 10 *s.* in *admission-money*, and 13 *l.* in *annual* payments. The true value is, over and above the *admission-money* just mentioned, an *annual payment* of 30 *l.* 17 *s.* (interest reckoned at 4 *per cent.*) or an *annual payment* of 36 *l.* 15 *s.* (interest reckoned at 3 *per cent.*)— The *London Union Society for the comfortable support of aged members* promises an annuity of no less than 50 guineas for life, after 50, to a person now 40 for 40 *l.* 10 *s.* in admission-money, and 7 *l.* in annual payments.

The

The *Amicable Society of Annuitants for the benefit of age*, promises an annuity of 26*l*. per annum, for life, to a person now 40, after attaining to 50, for 28 *l*. 16 *s*. in *admission-money*, and 6 *l*. in *annual payments*.—The true value of this annuity is 28 *l*. 16 *s*. in *admission-money*, and 17 *l*. 8 *s*. in *annual payments*, (interest supposed at 4 *per cent*.); or the same sum in *admission-money*, and 20 *l*. 18 *s*. in *annual payments*, interest supposed at 3 *per cent*.

The *Provident Society for the benefit of age* promises an annuity of 25 *l*. to a person now 40, after attaining to 50, for 34 guineas in admission-money, and eight guineas in annual payments. The true value is, 34 guineas in *admission-money*, and 15 *l*. 12 *s*. in *annual* payments, interest at 4 *per cent*.; or, the same sum in *admission-money*, and 19 *l*. in *annual* payments, interest being at 3 *per cent*. (*a*).

But I will not tire the reader, by going, in this manner, thro' the schemes of all these societies. The contrivers of them, it is certain, can know nothing of the principles on which the rule in Quest. VI. and the demonstration of it in the *Appendix* is founded; and, therefore, if unwilling to be guided by the authority of mathematicians, it may not be possi-

---

(*a*). The account here given of the terms on which a person whose age is 40, is admitted into these societies, I have taken from their printed Tables as they stood at the end of the year 1770.—In the younger ages, the deficiencies are greater.

ble to convince them of their mistakes. I will, however, offer to them the following demonstration, which will be understood, without difficulty, by every one who knows how to compute (*a*) the increase of money at compound interest.

The value of a life at 50, (interest being at 4 *per cent.*) is 11½ years purchase by Table VI. For an annuity, therefore, of 44 *l. per annum* for life, to be enjoyed by a person at this age, 498 *l.* ought to be given. *Two in three* of a number of persons at the age of 32 will, (by Tables III, IV, and V,) live to 50; and therefore, in order to be able to pay an annuity to them of 44 *l.* for life, after 50, the money now advanced by every *three*, ought to be such as will, in consequence of being laid up to be improved, increase in 18 years to double 498 *l.* or to 996 *l.*—From the preceding Table it may be learnt, that the money which ought to be advanced by every single person is 165 *l.* or by *three* persons 495 *l.* and this, in 18 years, will double itself, or increase to just the sum that will then be the value of the annuities to be paid.—But the money required in this case by the *Laudable Society*, is 14 *l.* 11 *s.* 9 *d.* from each member at admission, besides an *annual* payment of 4 *l.* The admission-money, therefore, of two members, being 29 *l.* 3 *s.* 6 *d.*

(*a*). The easiest method of doing this, is taught in the rules annexed to the Tables in the APPENDIX.

may be increased to twice this sum, or to 58 *l.* 7 *s.* An annual payment of 4 *l.* for 18 years will, if perfectly improved at 4 *per cent.* compound interest, increase to 102 *l*; and two such annual payments will increase to 204 *l.*

The whole pay, therefore, of *two* members will produce at the end of 18 years 262 *l.* 7 *s.*—A third part, I have said, will die without attaining to 50, and these will live one with another 9 years. An annuity of 4 *l.* for this time, will produce a capital of 42 *l.* 6 *s*; and this capital improved for nine years more will increase to 60 *l.* The whole profit, therefore, from the member who will die is, his admission-money doubled, and added to 60 *l.* or 89 *l.* 3 *s.* 6 *d.* And this sum added to 262 *l.* 7 *s.* makes 351 *l.* 10 *s.* 6 *d.* the *whole* money with which the society can be provided, at the end of 18 years, to bear the expence of *two* life-annuities, worth together 996 *l.*

By a similar computation it may be found, that the improvement of money at only 3 *per cent.* will *sink* the *former* sum to 324 *l.* at the same time that the value of the *annuities* will be *raised* to 1100 *l.*

The deficiencies in the schemes of most of the other societies, are no less considerable (*a*).—What confusion then must they pro-

---

(*a*) Some of these societies tell us, that the payments on admission shall increase, as the number of members in-

produce fome time or other? How barbarous is it thus to draw money from the public by pro-

increafes; and they have practifed on this rule juft as if the value of an annuity was nothing determinate in itfelf, but depended on the number of perfons who have been purchafers. But the true defign may perhaps be, to quicken the public in their applications.

Should any of thefe focieties, fenfible of their miftakes, refolve to reform themfelves, they ought to confider, that this cannot be done by only obliging *future* members to pay the juft values of the annuities promifed them. All the *prefent* members muft likewife, befides raifing their payments, make compenfation for what they have hitherto paid too little; and this compenfation is to be calculated in the following manner.—" Find the whole
" amount to the prefent time of the payments which have
" been made. Subtract this from the whole amount of
" the payments which *fhould* have been made; and the
" *remainder* will be the compenfation required."

EXAMPLE. In the *Laudable Society of Annuitants*, the condition of a title to 44 *l.* per annum for life, after 50, to a perfon at the age of 40, was, 4 years ago, 34 *l.* 17 *s.* in admiffion-money, befides an annual payment of 6 *l.* 14 *s.* 'till he attained to 50.—The admiffion-money will, (reckoning compound intereft at 3 *per cent.*) amount in four years to 39 *l.* 4 *s.* and the annual payment to 28 *l.* The whole amount, therefore, of the payments of a member, admitted 4 years ago, is 67 *l.* 4 *s.*—But the value of the annuity was 37 *l.* 4 *s.* in annual payments, befides 34 *l.* 17 *s.* in admiffion money; and thefe payments, during the 4 years, would have amounted to 195 *l.* The difference, therefore, between thefe two amounts, or 127 *l.* 16 *s.* is the *compenfation* which fuch member ought to pay; and if he continues a member without paying it, (befides raifing his annual contribution to 37 *l.* 4 *s.*) he muft either lofe his annuity, or owe it to injuftice.

I have taken intereft here at 3 *per cent.* becaufe I think thefe focieties cannot reafonably depend on always improving the money they receive at a higher rate.

Since

promises of advantages that *cannot* be obtained? Have we not already suffered too much by *bubbles?* (*a*)

I do not, however, mean to condemn all institutions of this kind. They may be very useful, if the full values are taken, and proper care is used in the *improvement* of money. Interest, in these cases, ought not to be reckoned higher than 3 *per cent.* and, supposing money improved at this rate, a person, for a single payment of 50 *l.* before he is 40, might be entitled to a life-annuity of 10 guineas *after* 55; or, if he chuses it, to a life-annuity of 17 *l. after* 60. But if he pays the same sum before he is 34, he might be entitled to a life-annuity of 14 *l, after* 55, or 22 *l. after* 60. 25 *l.* might purchase for him *half* these annuities; and 100 *l.* double.

Since I writ the above, I have found, that the admission-money required by this society has lately received another advance. At the age of 40, in particular, it is advanced to 108 *l.* 7 *s.*—when they have further either advanced the admission-money to *double* this sum, or *tripled* the annual payments, they will be *almost* right with respect to this particular age, provided the *compensation money*, just mentioned, has been paid.

These societies, tho' their plans are so insufficient, may, after beginning their payments to annuitants, continue them 15, or, perhaps, 20 years; but it will be by robbing all the younger members.

(*a*) See a farther account of these societies at the end of the SUPPLEMENT.

A society or office that would go on this plan, might do great service. Persons in the lower stations of life might be brought to a habit of industry, in the beginning of life, by striving to get 25 *l*. or 50 *l*. beforehand in order to purchase such annuities, and thus to make provisions for themselves in the more advanced parts of life, when they will be incapable of labour.

There are now established in *Holland* some institutions of this kind.—Any poor persons there, I am informed, who can, before they attain to a particular age, lay up 50 *l*. may make use of it in buying for themselves a right to be admitted, when 50, or at any time afterwards, to houses prepared on purpose, for providing them with all the conveniencies of lodging and board. This is an excellent institution; and I wish there was some imitation of it in this kingdom.

Considerable profits would, in this case, be received, from the payments of *some* who would chuse to *delay* going into such houses; and of others who would grow rich enough to be above them.

It is proper to observe here, that institutions of this kind would furnish one of the *safest* ways of providing for widows.—A married man might, by paying 100 *l*. before his wife attained to 40, entitle her, after 55, or 60, to a life-annuity of 21 *l*. or 34 *l*. Or, by paying the same sum before she attained to 34, he might entitle her, after the same ages, to a life-annuity of 28 *l*. or 44 *l*. (*a*); and in this case he would have a chance of sharing himself in the benefit of the annuity.

I have called this the *safest* way of providing for widows, because attended with none of the dangers arising from disproportion of age between men and their wives, and from the admission of persons labouring under concealed distempers.

I cannot conclude this Section, without mentioning the following plan of a provision for Old Age.

Let 13 guineas be given as *entrance-money*; and let besides 1 *l*. 2 *l*. 3 *l*. 4 *l*. &c. be given at the beginning of the first, 2d, 3d, 4th, &c. years, as the payments for these years respectively; and let the last payment be 16 *l*. at the beginning of the 16th year. All these payments put together will, according to the probabilities of life in the 3d, 4th, and 5th Tables, (interest being at 4 *per cent*.) entitle a person, whose age was 40 when he begun them, to an annuity, after 15 years, beginning with 15 *l*. and increasing at the rate of 1 *l*. every year, 'till at the end of 15 years

---

(*a*) The same payment before 30, would entitle to an annuity of 22 *l*. after 50.

more, or (*a*) when he has attained to 70, it becomes a standing annuity of 30 *l.* for the remainder of his life.

If the addition of three guineas is made to the *entrance-money*, for every year that any life between 30 and 40 falls short of 40, the value will be obtained nearly, of the same annuity to be enjoyed by that life, after the same number of years, and increasing in the same manner, 'till, in 30 years, it becomes *stationary* and *double.*—This plan is particularly inviting, as it makes the *largest* payments become due, when the *near* approach of the annuity renders the encouragement to them *greatest*; and as, likewise, the annuity is to increase continually with age, 'till it comes to be highest (*b*), when life is most in the decline,

---

(*a*) According to the probabilities of life in the *London* Table, this annuity should be greater.—A *Theorem* for finding what the annuity ought to be in these cases, is given in the Appendix, Note (I).

(*b*) The lower part of mankind are objects of particular compassion, when rendered incapable, by accident, sickness, or age, of earning their subsistence. This has given rise to many very useful societies among them, for granting relief to one another, out of little funds supplied by *weekly* contributions. A society of this kind, formed on the following plan, would probably thrive, and might, on some accounts, be even more useful than the institutions in *Holland*, mentioned in p. 116.

Let the society, at its first establishment, consist of 100 persons, all between 30 and 40; and whose mean age may

cline, and when therefore it will be most useful.—It is further a recommendation of this plan, that less depends in it on the *improvement* of money than in most other plans.—But I must leave these hints to be pursued by others.

may therefore be reckoned 36; and let it be supposed to be always kept up to this number, by the admission of new members, between the ages of 30 and 40, as old members die off. Let the contribution of each member be four-pence *per* week, making, from the whole body, an annual contribution of 85 *l.* 17 *s.*—Let it be further supposed, that seven of them will fall every year into disorders, that shall incapacitate them for seven weeks.—30 *l.* 12 *s.* of the annual contribution will be just sufficient to enable the society to grant to each of these 12 *s. per* week, during their illnesses. And the remaining 55 *l. per annum*, laid up and *carefully* improved, at $3\frac{1}{4}$ *per cent.* will increase to a capital that shall be sufficient, according to the chances of life in Tables III, IV, and V, to enable the society to pay to every member, *after* attaining to 67 years of age, or *upon* entering his 68th year, an annuity, beginning with 5 *l.* and increasing at the rate of 1 *l.* every year for seven years, 'till, at the age of 75, it came to be a standing annuity of 12 *l.* for the remainder of life.

Were such a society to make its contribution *sevenpence per* week, an allowance of 15 *s.* might be made, on the same suppositions, to every member during sickness; besides the payment of an annuity beginning with 5 *l.* when a member entered his 64th year, and increasing for 15 years, 'till, at 79, it became fixed for the remainder of life at 20 *l.*

If the probabilities of life are lower among the labouring poor, than among the generality of mankind, this plan will be so much the more sure of succeeding.

SECT. V.

Of the Amicable Society for a perpetual Assurance Office: And the Society for Equitable Assurances on Lives and Survivorships.

THE 10th Problem has been given, with a particular view to the corporation of the *Amicable Society*, for a perpetual Assurance-Office on single lives, kept in *Serjeant's-Inn*. This society was established in 1706, and is the only one I am acquainted with, which has stood any considerable trial from time and experience. The annual payment of each member used to be 6*l*. 4*s*. payable quarterly; but it has been lately reduced to 5*l*. The whole annual income, hence arising, is equally divided among the *nominees*, or heirs of such members as die every year; and this renders the dividends among the *nominees* in *different* years, more or less, according to the number of members who have happened to die in those years. But the society now engages, that the dividends shall not be *less* than 150*l*. to each claimant, though they may be *more*.—None are admitted whose ages are *greater* than 45, or *less* than 12; nor is there any difference of contribution allowed on account of difference of age.

for assuring Lives.

This society has, I doubt not, been very useful to the public; and its plan is such, that it cannot well fail to *continue* to be so. It might, however, certainly have been much more useful, had it gone from the first on a different plan. It is obvious, that regulating the dividends among the *nominees* by the number of members who die every year, is not *equitable*; because it makes the benefit which a member is to receive to depend, not on the value of his contribution, but on a *contingency*; that is, the number of members that shall happen to die the same year with him. This regulation must also have been disadvantageous to the society; as will appear from the following account of the natural progress of the affairs of such a society, when established on a right plan.

Suppose a *thousand* persons, whose common age is 36, to form themselves into a society for the purpose of *assuring* a particular sum at their deaths, to such persons as they shall name, in consideration of a particular annual-contribution to be continued during their lives. Suppose the annual contribution to be 5 *l*. and the first payment (*a*) to be made immediately. Suppose, likewise, the original number of the society to be constantly kept up by the admission of new members,

---

(*a*) Such payments, it has been shewn, Quest. VIII. p. 28, are better than any *half* yearly or *quarterly* payments, and at the same time they save some trouble.

at 36 years of age, in the room of such as die.—In Quest. X. p. 33, it appears, that an annual payment, beginning immediately, of 5*l*. during a life now at the age of 36, should entitle, at the failure of such a life, to 172 *l*. reckoning interest at 4 *per cent*. and taking Mr. *De Moivre's* valuation of lives.—A *thousand* persons, all 36 years of age, will die off at the rate of 20 every year. The disbursements, therefore, of such a society will be, the first year, 20 times 172 *l*. or 3440 *l*. and its income will be 5000 *l*. It will, therefore, at the end of the year, have a surplus of 1560 *l*. to put to interest.—In consequence of the yearly accessions to supply vacancies, the number dying annually will be always increasing after the first year. In 50 years it will attain to a *maximum*; and then, the affairs of the society will become *stationary*, and the number dying annually will be 40, and its annual expence will be 6,880 *l*. exceeding the annual contribution, 1,880 *l*. But, in the mean time, by improving its surplus monies, it will have raised a capital equal to this excess, and, consequently, its affairs will be fixed on a firm basis for all subsequent times.

Suppose now, that such a society, at its establishment, should resolve to divide its whole yearly income among the *nominees* of deceased members. The effect of this would be, be, that no capital could be raised; that the dividends payable to *nominees* would diminish continually, 'till, at the time that the greatest number of members came to die *annually*, or at the end of 50 years, they would be reduced to half; and all claimants, after this period, receive too little, because the first claimants had received too much (*a*).

At the time of the institution of the *Amicable Corporation*, the interest of money was at 6 *per cent*. and, as they admit all between 12 and 45, the mean age of admission cannot probably be so great as 36. It appears, therefore, that had they avoided the error now mentioned, and gone from the first on

---

(*a*) The reverse of this will take place, if such a society *begins* with admitting all at all ages, and afterwards changes its plan, and *limits* the age of admission. In this case, the number of *yearly deaths* will be *greatest* at first, and the *dividends smallest*. In consequence of altering its plan, the *yearly deaths* will lessen gradually, and the *dividends* rise; but in time *both* would return again to their original state.

The following facts incline me to suspect, that this remark may be applicable to the *Amicable Corporation*.

First. In their *original charter*, as it is given in their printed abstracts, there is no limitation of age mentioned; but 31 years afterwards, I find a bye-law made against admitting any person who should be above the age of 45, or under 12.—Secondly. In their printed advertisements in 1770, it is said, that in 59 years they had paid, among 3643 claimants, 378,184 *l*. from whence it follows, that tho' the average of their dividends, for the last 17 years, has been 154 *l*. the same average, for 59 years, is only 104 *l*.

the

the plan I have described; they might have all along paid to each *nominee* 172 *l.* besides raising a capital much greater, in proportion to the number of members, than that I have specified; by the help of the excess of their annual payments above 5 *l.* and some other advantages which they have enjoyed (*a*). Indeed, I cannot doubt but that, with these advantages, they might, before this time, have found themselves able to pay at least 200 *l.* to each *nominee*; and at the same time restricted themselves, as they now do, to an annual payment of 5 *l.* (*b*).

I have already mentioned one instance in which the plan of this society is not equitable. Another instance of this is, their requiring the same payments from all persons under 45, without regarding the differences of their ages; whereas, the annual payments of a person admitted at 45, ought to be double the annual payment of a person admitted at 12.

(*a*) A surplus from a *thousand* members of only *five shillings per annum*, duly improved, at 4 *per cent.* would, in 41 years, produce a capital of 25,000 *l.*

(*b*) It should be remembered, that all this is said on the supposition, that proper care has been taken to keep out unhealthy persons; and that the probabilities of life among the members of this society, are the same with those in the 3d, 4th, and 5th Tables, in the *Appendix.*

Further.

for assuring Lives.

Further. The plan of this society is so narrow, as to confine its usefulness too much. It can be of no service to any person whose age exceeds 45. It is, likewise, far from being properly adapted to the circumstances of persons, who want to make assurances on their lives, for only short terms of years.—Thus; the true value of the assurance of 150 *l.* for 10 years, on the life of a person whose age is 30, is, by Quest. XIV. (interest being at 3 *per cent.*) 2 *l.* 13 *s.* in annual payments, for 10 years, to begin at the end of the first year; and subject to failure when the life fails. But such an assurance could not be made, in this society, without an annual payment of 5 *l.*—Neither is the plan of this society at all adapted to the circumstances of persons, who want to make assurances on particular survivorships.—For example. A person possessed of an estate, or salary, which must be lost with his life, has a person dependent upon him, for whom he desires to secure a sum of money, payable at his death. But, he desires this only as a provision against the danger of his dying *first*, and leaving a wife, or a parent, without support. In these circumstances, he enters himself into this society; and by an annual payment of 5 *l.* entitles his *nominee* to 150 *l.* In a few years, perhaps, his *nominee* happens to die; and, having then lost the benefit he had in view, he determines to forfeit his former payments, and

and to withdraw from the society. In this way, probably, this society muſt have gained ſome advantages. But the right method would have been, to have taken from ſuch a perſon the true value of the ſum aſſured, " on the ſuppoſition of non-payment, pro- " vided he ſhould ſurvive." In this way he would have choſen to contract with the ſociety; and had he done this, he would have paid for the *aſſurance*, (ſuppoſing intereſt at 3 *per cent*. his age 30, the age of his *nominee* 30, and the probabilities of life as in the 3d, 4th, and 5th Tables) 3 *l*. 8 *s*. (*a*) in annual payments, to begin immediately, and to be continued during the *joint* continuance of his own life, and the life of his *nominee*.

All theſe objections are removed by the plan of the Society kept in *Nicholas-Lane, Lombard-Street*, which has juſtly ſtiled itſelf the *Society for Equitable Aſſurances on Lives and Survivorſhips*. This Society, if due care is taken, may prove a very great public benefit. It was founded, in conſequence of

---

(*a*) The value of 150 *l*. payable at the death of a perſon, aged 30, *provided* he ſurvives another perſon of the ſame age, is, by Queſt. XI. Chap. I. *l*. 45.65; and this value divided by 13.43, (the value increaſed by unity, of two joint lives both 30) gives; *l*. 3.4, or 3 *l*. 8 *s*.— The value of the ſame reverſion, according to the probabilities of life in *London*, is, *l*. 49.19, in one payment; and 4.16, in *annual* payments, during the joint lives, the firſt payment to be made immediately.

proposals which had been made, and lectures, recommending such a design, which had been read by Mr. *Dodson*, the author of the *Mathematical Repository*. It assures any sums or reversionary annuities on any lives, for any number of years, as well as for the whole continuance of the lives, at rates settled by particular calculation; and in any manner that may be best adapted to the views of the persons assured. That is; either by making the assured sums payable *certainly* at the failure of any given lives; or on *condition* of survivorship; and also, either by taking the price of the assurance in *one present payment*; or in *annual payments*, during any single or joint lives, or any terms less than the whole continuance of the lives.—In short; the plan of this society is so extensive, and so important, that I cannot satisfy my own mind, without offering to the gentlemen concerned in the direction of it, the following observations, hoping they will not think them impertinent or improper.

First. They should consider what distress would arise from the failure of such a scheme in any future time; and what dangers there are, which ought to be carefully guarded against in order to secure success. I have already more than once observed, that those persons will be most for flying to these establishments, who have feeble constitutions,

or

or are subject to distempers, which they know render their lives particularly precarious; and it is to be feared, that no caution will be sufficient to prevent all danger from hence.

Again. In matters of chance, it is impossible to say, that an unfavourable run of events will not come, which may hurt the best contrived scheme. The calculations only determine probabilities; and, agreeably to these, it may be depended on, that events will happen on the whole. But at particular periods, and in particular instances, great deviations will often happen; and these deviations, at the commencement of a scheme, must prove either very favourable, or very unfavourable.

But further. The calculations suppose, that all the monies received are put out immediately to accumulate at compound interest. They make no allowance for losses, or for any of the expences attending management. On these accounts, the payments to a society of this kind, ought to be more than the calculations will warrant. The interest of money ought to be reckoned low; and such Tables of Observation used as give the highest values. Mr. *Dodson*, I find, has paid due attention to all this, by reckoning interest, in his calculations for this society, at 3 *per cent*. and taking the lowest of all the known probabilities of life, or those deduced

from

from the *London* bills of mortality (*a*). There is, besides, a liberty provided of making a call on all the members, in case of any particular emergency. It is, therefore, highly probable, that this society (provided too much money is not spent in management) must be secure. The last expedient, however, would be a very disagreeable one, should there be ever any occasion for having recourse to it; and, in order to guard still more effectually against danger, it would not, I think, be amiss to charge a profit of 3 or 4 *per cent*. on all the payments.—Should the consequence of this prove, that in some future period the society shall find itself possessed of too large a capital, the harm will be trifling, and future members will reap the advantage. But this leads me to repeat an observation of particular consequence.

As this society is guided in every instance, by strict calculation, it is not to be expected

---

(*a*) It ought, however, to be remembered here, that in selling life-annuities to commence either immediately, or after given terms; and also in some other cases, the values come out *less* in consequence of *lower* probabilities of life. Would it, in *such* instances, be taking an unfair advantage, to estimate the values by the 3d, 4th, or 5th Table in the Appendix, rather than the *London* Table?— Thus; was the society to sell 20 *l. per annum*, for life, to a person now 30, after attaining to 50, the value, according to Dr. *Halley*'s Table, would, reckoning interest at 3 *per cent*. be 90 *l*. in a single payment; but, according to the *London* Table, the value would be only 70 *l*.

K   that

that it can meet with any difficulties for many years; becaufe, not 'till the end of many years after it has acquired its *maximum* of members, will the *maximum* of yearly claimants and annuitants come upon it? Should it, therefore, thro' inattention to this remark, and the encouragement arifing from the poffeffion of a large furplus, be led to check or ftop the increafe of its ftock by enlarging its dividends too foon, the confequences might prove pernicious.

Again. I would obferve, that it is of great importance to the fafety of fuch a fociety, that its affairs fhould be under the infpection of able mathematicians. Melancholy experience fhews, that none but mathematicians are qualified for forming and conducting fchemes of this kind.—In fhort; dangerous miftakes may fometimes be committed, if the affairs of fuch a fociety are not managed frugally, carefully, and prudently. One inftance of this I cannot avoid mentioning.

A perfon, who defires to affure a particular fum, to be paid at the failure of his life, on condition of the furvivorfhip of another life, may chufe to pay the value in annual contributions during the continuance of his own fingle life, rather than during the continuance of the joint lives, becaufe the annual contributions, in this cafe, ought to be much lefs. But a fociety that would practife fuch a method of *affurance* would hurt itfelf;

itself; for, as soon as the life, on whose survivorship the assurance depends, is extinct, the person assured, if then living, would have no longer any benefit in view; and, therefore, would make his payments with reluctance, and in time, perhaps, entirely withdraw them; the consequence of which would be, that the society would suffer a loss by being deprived of the just value of the expectation it had granted. The plan of a society ought always to be such, as that the losses arising from discontinuance of payment, should fall on the purchaser, and never on the society.

I must not forget to add, that it is necessary, that such a society should be furnished with as complete a set of Tables as possible. This will render the business of the society much more easy; and also much more capable of being conducted by persons unskilled in mathematics. It will also contribute much to its *safety*. For in all cases to which Tables can be extended, there would be no occasion for employing any calculators; and, consequently, a danger would be prevented to which, tho' it is not *now*, it may *hereafter* be exposed; I mean, the danger of happening to trust unskilful, or careless calculators.—Mr. *Dodson*, I find, has furnished this society with some important Tables; and his skill was such, that there is no reason to doubt, but they may be depended on. They have also others which, I believe, are safe and accurate. But there are some still wanting which should be supplied; and all should be subjected to the examination of the best judges, and afterwards published; together with a minute account of the principles assumed, and the method taken in composing them. Such a publication would be a valuable addition to this part of science; and it would also be the means of increasing and establishing the credit of the society.

In Questions 4th, 6th, 10th, 11th, 14th, 15th, and 16th, I have, with a particular view to this society, given rules, by which may be formed every Table it can want, for shewing the values of assurances on the *whole duration*, or any *terms*, of any *one* or *two* lives, in all possible cases; and nothing but care and attention can be necessary to enable any good arithmetician to calculate from them. Perhaps, this may be as much business as any one society should undertake. Rules, however, for finding the values of *assurances*, in most cases, where the whole duration of any *three* lives is concerned, may be found in Mr. *Simpson*'s Select Exercises, from page 299 to p. 307; and it is not possible they should follow a better guide.

CHAP. III.

Of PUBLIC CREDIT, *and the* NATIONAL DEBT.

THE *National Debt* is a subject in which the public is deeply interested. Some observations have occurred to me upon it, which I think important; and for this reason, though foreign to my chief purpose in this work, I cannot help here begging leave to offer them to public attention.

The practice of raising the necessary supplies for every national service, by borrowing money on interest, to be continued till the principal is discharged, must be in the highest degree detrimental to a kingdom, unless a plan is settled, for putting its debts into a regular and certain course of payment. When this is not done, a kingdom, by such a practice, obliges itself to return for every sum it borrows infinitely greater sums; and, for the sake of a present advantage, subjects itself to a burden which must be always growing heavier and heavier, 'till it becomes insupportable.

This seems to be now the very state of this nation. At the REVOLUTION, an æra in other respects truly glorious, the practice I have mentioned begun. Ever since, the public debt has been increasing fast, and every new war has added much more to it, than was taken from it, during the preceding period of peace. In the year 1700, it was 16 millions. In 1715, it was 55 millions. A peace, which continued 'till 1740, sunk it to 47 millions; but the succeeding war increased it to 78 millions; and the next peace sunk it no lower than 72 millions. In the *last* war it rose to 148 millions. During a peace which has lasted now 10 years, it has been reduced to near 138 millions: And at a sum not much less than this, it will, perhaps, be found at the commencement of another war, which may possibly raise it to 200 millions.—One cannot reflect on this without terror.—No resources can be sufficient to support a kingdom long in such a course. 'Tis obvious, that the consequence of accumulating debts so rapidly; and of mortgaging posterity, and funding for eternity, in order to pay the interest of them; must in the end prove destructive. Rather than go on in this way, it is absolutely necessary, that no money should be borrowed, except on annuities, which are to terminate within a given period. Were this practised, there would be a LIMIT beyond which the national debts could not increase; and time would do that *necessarily*

for

for the public, which, if trusted to the œconomy of the conductors of its affairs, might possibly *never* be done.

This, therefore, is one of the proposals to which, on this occasion, I wish I could engage attention.—I am sensible, indeed, that the *present* burdens of the state would, in this case, be increased, in consequence of the greater present interest, that would be necessary to be given for money. But I do not consider this as an objection of any weight. For let the annuity be an annuity for a 100 years. Such an annuity is, to the present views of men, nearly the same with an annuity for ever; and it is also nearly the same in calculation, its value at 4 *per cent.* being 24½ years purchase, and therefore only half a year's purchase less than the value of a *perpetuity*. Supposing, therefore, the public able to borrow money at 4 *per cent.* on annuities for ever, it ought not to give above 1 *s.* 7 *d. per cent.* more for money borrowed on annuities for 100 years: But should it be obliged to give a *quarter*, or even an *half per cent.* more(*a*), the additional burdens derived from hence, would

(*a*) These annuities might be kept 18 years without being much diminished in value; for, supposing interest at 4 *per cent.* an annuity for 82 years, is, within a 49th part, or 2 *l.* in 98 *l.* worth as much as an annuity for a 100 years.

Perhaps, in this way of raising money, it might be best to offer a higher interest at first, which should fall to a lower,

would not be such as could be very sensibly felt; and the advantages, arising from the necessary annihilation of the public debts by time, would abundantly overbalance them.

These advantages would be, indeed, unspeakably great. By such a method of raising money, the expence of one war would, in time, come to be always discharged, before a new war commenced; and it would be impossible, that a state should ever have upon it, at any one time, the expence of many wars; or any larger debts than could be contracted, within the limited period of the annuities: and, consequently, it would enjoy the invaluable privilege of being rendered, in some degree, independent of the management of its finances by ignorant or unfaithful servants.

I must add, that it is by no means necessary, that the limited period of the annuities should be so long as, I have mentioned, or 100 years: And that, at any time before the expiration of this period, the public might employ any surplus monies, in extinguishing part of the annuities, by purchasing them for itself at the market price; and thus it might aid the operations of time, and keep its debts within any bounds, that its interest rendered

lower, at the end of given intervals. Thus, tho' $4\frac{1}{2}$ for 100 years is equal in value to 5 *per cent.* for 17 years, and after that 4 *per cent.* for 83 years, yet the latter might appear more inviting.

neceſſary. Our government has, I know, in ſome inſtances adopted the plan now propoſed; but it is to be wiſhed that, inſtead of retracting (*a*) it, as was once done, it had been carried much further.

I am, however, far from intending to recommend this plan as the beſt a ſtate can purſue. There is another method of gaining the ſame end, which is, on many accounts, preferable to it. I mean, " by providing an
" annual ſaving, to be applied invariably,
" together with the intereſt of all the ſums
" redeemed by it, to the purpoſe of diſcharg-
" ing the public debts: Or, in other words,
" by the eſtabliſhment of a permanent SINK-
" ING FUND."

It is well known, that this plan has been alſo adopted by our government; but, tho' capable of producing the *greateſt* effects in the *eaſieſt* and *ſureſt* manner, it has never been carried into execution. It will abundantly appear from what follows, that this obſervation is juſt.

Suppoſe the annual ſaving to be 100,000 *l*. This ſum, applied now to diſcharge an equal debt, bearing intereſt at 4 *per cent*, will transfer to the public, from its creditors, an an-

---

(*a*) In the year 1720, the nation was put to the expence of above three millions, in order to reduce ſeveral long and ſhort annuities then ſubſiſting, to redeemable *perpetuities*.

nuity of 4,000 *l.* At the end of a year, then, there would be a saving of 104,000 *l.* which would transfer to the public another annuity of 4,160 *l.* and make the saving, at the end of two years, to be 108,160 *l.*—Thus, the original fund would go on increasing, at the same rate with money improved at 4 *per cent.* compound interest.—At the end of three years it would be 112,486 *l.* At the end of 18 years, 202,581 *l.* Of 36 years, 410,393 *l.* and of 95 years (*a*), 4,151,138 *l.*—At the end of 93 years, then, the nation might be eased of above 4 millions *per annum* in taxes; and above 100 millions of its debts would be discharged, gradually and insensibly, at no greater expence than 100,000 *l. per annum*; and, without interfering with any of the resources of government; or making any other difference, than causing *funds* to be engaged for a course of time to the *public*, which would have been otherwise necessarily engaged to its *creditors*, and which, therefore, must have been entirely useless to it.

It is an observation that deserves particular attention here, that, on this plan, it will be of less importance to a state what interest it is obliged to give for money: For the higher the interest, the sooner will such a fund pay off the principal. Thus; a 100 millions borrowed at 8 *per cent.* and bearing an an-

---

(*a*) See the Questions annexed to the Tables in the Appendix.

nual intereſt of eight millions, would be paid off by a fund, producing annually 100,000 *l*. in 56 years; that is, in 39 years leſs time, than if the ſame money had been borrowed at 4 *per cent.* (a).

It follows from hence, that reductions of intereſt would, on this plan, be no great advantage to a ſtate. They would, indeed, lighten its preſent burdens; but this advantage would be, in ſome meaſure, balanced

---

(a) What is here ſaid, ſuppoſes the *ſame* fund applied to the diſcharge of debts bearing *different* intereſts. If different funds are applied, bearing to one another the ſame proportion with the intereſts of the debts which they are to diſcharge, the benefit derived from borrowing on lower rather than higher intereſts, will be reduced to almoſt nothing; for the diſburſements of the public on account of all equal loans, will, in this caſe, be very nearly the ſame.

The following example will explain and demonſtrate this:

Let a million be borrowed at 3 *per cent.* and let a fund be charged with it, bringing in *ſix ſhillings per cent. per ann.* more than the intereſt; or 33,000 *l*. inſtead of 30,000 *l*. *per ann.* This *ſurplus, unalienably* applied, together with all the intereſts diſengaged by it, will annihilate the *principal* in 81 years, as may be gathered from Queſtion V. in the *Appendix*. And the diſburſements, on account of the loan, will be 81 multiplied by 33,000 *l*. that is, 2.673,000 *l*. Let us ſuppoſe again, a million borrowed at 6 *per cent.* and let a fund be charged with it, producing a ſurplus of *twelve ſhillings per cent. per ann.* ſuch a fund, beſides paying the intereſt, will diſcharge the *principal* in 41 years; and the diſburſements, on account of the loan, will be 66,000 *l*. multiplied by 41; that is, 2.706,000 *l*. or nearly the ſame with the diſburſements on account of an equal loan at 3 *per cent.*

by

by the addition which would be made to its *future* burdens, in confequence of the longer time, during which it would be neceffary to bear them.—I mean this on the fuppofition, that the favings produced by reductions of intereft, are immediately applied to the relief of the ftate, by annihilating taxes equivalent to them. But if that is not the cafe; and if, likewife, there is either no plan eftablifhed for putting the public debts into a certain courfe of payment, or it is not faithfully carried into execution; in thefe circumftances, reductions of intereft may prove hurtful. For, firft, They would only furnifh with more money for fupplying the deficiencies arifing from profufion and bad management. And, fecondly, As, in fuch circumftances, they would only *retard*, and not *prevent* the increafe of the burdens occafioned by the public debts, a period would come when the affairs of the ftate would get to a *crifis*; and at fuch a period, its danger would be increafed, in proportion to the reductions of intereft that had been made.

In order to underftand this; let us fuppofe that a debt, bearing an annual intereft of five millions, is the whole debt, which a ftate can bear without being fo much oppreft as to be near finking. Let it, however, be fuppofed to have ftill fome laft refources left, which may enable it to bear, for 23 years to come, this load, together with every additional

tional load, which, during this time, may be neceſſary to be thrown upon it.—Let it further be ſuppoſed, that at this time, the ſtate, urged by the fear of an approaching bankruptcy, reſolves upon entering into ſome effectual meaſures for preſerving itſelf.——— Certain it is, that in ſuch circumſtances, no meaſure *ſo* effectual can be purſued, as the eſtabliſhment of a *ſinking fund*, and ſuch a faithful application of it as I have explained. Let that then be the meaſure entered upon; and let the ſtate be ſuppoſed capable of providing a fund, producing a million annually. If all the debts bear intereſt at 6 *per cent.* this fund would pay off three-fifths of them, within the time I have mentioned; or, in 23 years; and the ſtate might be ſaved. But if, in conſequence of reductions, they bear intereſt at no more than 3 *per cent.* the ſame fund would not give the ſame relief, in leſs than *double* that time; and, therefore, a bankruptcy might prove unavoidable.

I wiſh I could think, that there is nothing in this repreſentation, that can be applied to the preſent ſtate of this nation. The intereſt of the public debts has been reduced, at different periods, from 6 to 5, from 5 to 4, and from 4 to 3 *per cent.*; but ſtill they have grown with rapidity; and we now ſee ourſelves overloaded, and in no way of gaining relief. Had there been no reductions of intereſt, we ſhould, indeed, have been in the same condition sooner; but, we might have been relieved also sooner, and with less difficulty and danger.

In short. Reductions of interest are advantageous chiefly when made to gain additions to such a *sinking fund* as I have described.——When made with other views, they are only *palliatives* which give *present* relief by increasing *future* danger; or *expedients* which *postpone* a public bankruptcy, by rendering it a calamity more *unavoidable* and *dreadful*. As managed therefore, among us, they have been indeed the effects of too narrow a policy, and deserve none of the *encomiums* which have been bestowed upon them.——The preceding observations prove this sufficiently; but there is one farther proof of it which I cannot help mentioning.—Suppose 200,000 *l. per ann.* to have been gained in 1716, by the reduction which was then made of the 6 *per cents.* to 5 *per cents*; or, in other words, by saving 1 *per cent. per ann.* on a capital of 20 millions. This saving, in consequence of being applied unalienably in the manner I have represented, to the payment of the public debts, would, in 37 years, have discharged a debt of 20.325,000 *l.* bearing 5 *per cent.* interest. But if applied every year to current services, in order to avoid levying new money, the benefit derived from it in the same period, would be 37 times 200,000 *l.* or 7.400,000 *l.*

but

but at the fame time, a debt would have been continued of 20 millions, which muſt have been otherwiſe paid. The effect, therefore, in this caſe, of the reduction, would be to prevent an incumbrance on the public of 200,000 *l. per ann.* by leaving upon it an incumbrance of a million *per ann.* rendered more difficult and unlikely than ever to be removed.

But to return to the ſubject I have principally in view.

What I have ſaid implies, that a ſtate always diſcharges its debts, whatever intereſt they bear, by paying the original ſum borrowed. It may, perhaps, be imagined, that when a loan is under *par*, it may be diſcharged at a leſs expence. But this is by no means ſo practicable as it may ſeem; for it ſhould be conſidered, that a public loan, now under *par*, would not long keep ſo, after being put into a courſe of payment: And, for this reaſon, as a ſtate can never be obliged, in redeeming its debts, to pay *more* than the original ſum borrowed, ſo neither ought it to expect, in general, to be able to redeem them by paying *leſs*. I have ſaid, *in general*; for I am ſenſible, that at the beginning of the operations of a fund, when its produce is ſmall; and alſo, in a time of war, a ſtate might derive great advantages from the low price of its debts. And I am ſenſible alſo, that

that confiderable advantages might be derived from *lotteries* (*a*), in paying the public debts: But *lotteries* do great mifchief in a ftate, by foftering the deftructive fpirit of gaming. It is wretched policy to make them familiar, by recurring to them in the ordinary courfe of government. There are great occafions on which they may be neceffary, and for fuch occafions they fhould be referved.

The advantages of putting the public debts into fuch a courfe of payment as I have defcribed, are fcarcely to be imagined. It would give a vigour to public credit, which would enable a ftate always to borrow money eafily, and on the beft terms. And the encouragement to lenders might be always improved, without any inconvenience, by making every loan irredeemable, during the firft 20 or 30 years; for, there could feldom be any occafion, for beginning to difcharge any *one* loan fooner.

It might be eafily fhewn, that the faithful application, from the beginning of the year 1700, of only 200,000 *l.* annually, would long before this time, notwithftanding the

---

(*a*) Thus; 800,000 *l.* of the 3 *per cents.* at 87; or 1,000,000, at 70, might be redeemed with half a million of money, confifting of 50,000 lottery tickets at 10 *l.* each, real value; but capable of being fold at 14 *l.* as was done in fome of the laft lotteries.

reductions of intereſt, have cauſed above half the public funds to revert to the public, and paid off above 80 millions of its debts. The nation might, therefore, ſome years ago, have been eaſed of the greateſt part of the taxes with which it is loaded. The moſt important relief might have been given to its trade and manufactures; and it might now have been in much better circumſtances, than at the beginning of the laſt war; its credit firm; reſpected by foreign nations; dreaded by its enemies; and ready to puniſh any inſult that could be offered to it. The near view, likewiſe, of ſuch a period, during the courſe of the laſt war, would have given higher ſpirits to the nation, and encouraged it to bear the expence occaſioned by the war with more chearfulneſs, and to continue it with vigour for two or three years longer; the conſequence of which would, probably, have been, gaining a full indemnification from our enemies, and weakening them to ſuch a degree, as would have given us effectual ſecurity againſt them for many years to come.—A new account might alſo now have been begun; and another fund, not much more conſiderable, applied in the ſame way, would, in 60 or 70 years more, have paid, not only all that would have been now unpaid, but alſo, probably, a great proportion of ſuch further debts as must be contracted within this time (*a*). And thus, without any expence that could be sensibly felt, its debts, as soon as they began to grow heavy, might have been constantly reduced to a *half*, or a *third*; and not only all *danger*, but all considerable *inconvenience* from them prevented.

All I have now said, supposes a *single* fund with a *general* appropriation to the payment of the public debts. The same ends might be answered by *particular* funds, with small surplusses, appropriated to *particular* debts. In the wars of King *William* and Q. *Anne*, 6 *per cent*. interest was given for all loans. It would have been easy to have annexed to each loan a *fund* producing a *surplus* of 1 *l. per cent*. after paying the interest; and such a *surplus* would have been sufficient to annihilate the principal of every loan in 33 years. Had this plan been followed, the disengagement of the public funds, and the relief attending it, would have begun 50 years ago; and the debts contracted, during the reigns of King *William* and Queen *Anne*, would have been all cancelled near 20 years ago, without

---

(*a*) One of the properest objects of taxation in a state is *celibacy*. I doubt not, but that by a fund supplied only from hence, the end I have in view might have been easily accomplished; and, consequently, the very means of paying off the debts of the nation, rendered at the same time the means of increasing its chief strength, by promoting population in it.

any of that trouble, tumult, and diſtreſs, which have been occaſioned by reductions of intereſt; and by the various ſchemes which have been tried for leſſening the debts (*a*).—A fund, yielding 1 *l. per cent.* ſurplus, annexed to a loan at 5 *per cent*, would diſcharge the principal in 37 years (*b*). At 4 *per cent*, in 41 years. At 3 *per cent*, in 47 years.

Theſe obſervations relate only to what *might* have been the ſtate of the nation with reſpect to its debts, had a right plan been purſued from the firſt. But it will be aſked, What can be done with them as they *are?*— I wiſh I was able to give a more ſatisfactory anſwer to this enquiry. Every one muſt ſee our proſpect to be diſcouraging, and our ſtate hazardous. Some have thought, that a good method might be found out of diſcharging

---

(*a*) The ſums to be laid out would, in this caſe, be ſo ſmall at firſt, that it would be proper to employ them in purchaſing part of the loan to be annihilated at the prices in the public market; and this, as far as it can be carried, is the moſt eaſy and quiet and ſilent way poſſible of extinguiſhing the public debts.

(*b*) I have all along ſuppoſed the produce of the public funds to come in yearly. The truth is, that it comes in *half*-yearly; but this gives no advantage in the payment of the public debts worth taking into account. 1 *l. per annum*, together with its growing intereſt, at 4 *per cent.* taken *yearly* out of 100 *l.* will reduce it to nothing in 41 years; if taken *half*-yearly, it will annihilate the ſame capital only four months and 12 days ſooner. See the Queſtions annexed to the Tables in the *Appendix*.

the national debt; by life annuities. The following obfervations will fhew how vain an imagination this is.

Let us fuppofe, that 33,333,000 *l.* is to be paid off, by offering to the public creditors life-annuities, in lieu of their 3 *per cents*. A life at 60, fuppofing intereft at $3\frac{1}{2}$ *per cent.*, and the probabilities of life as in the *Breflaw*, *Norwich*, and *Northampton* Tables of Obfervation, is worth 9 years purchafe. A life at 30 is worth $15\frac{1}{2}$ years purchafe. Certainly, therefore, no fcheme of this kind would be fufficiently inviting, which did not offer 8 *per cent.* at an average, to all fubfcribers. Let us, however, fuppofe, that no more than $7\frac{1}{2}$ is given; and that there are 33333 fubfcribers, at 1000 *l.* ftock each, for which a life-annuity is to be granted of 75 *l.* or, for the whole ftock fubfcribed, two millions and a half. A million and a half extraordinary, therefore, muft be provided every year, towards paying thefe annuities.

Let us farther fuppofe that the fubfcribers are perfons between the ages of 30 and 60; and that the numbers of them, at all the intermediate ages, are in the fame proportions to one another, with the proportions of the living at thefe ages, as they exift in the world, or, as they are given in *Tables of Obfervation*. Let us again fuppofe, that as thefe *annuitants* die off, they are immediately replaced by others, who are continually offering

fering themselves at the same ages, and in the same proportional numbers at these ages, with those of the original subscribers at the time they subscribed; in consequence of which, the whole number of annuitants will be kept always the same. In these circumstances, it will be 30 years, at least, before a number will die off (*a*), equal to the whole number; that is, before 33 millions of debts will be annihilated. But had the extraordinary million and half provided for paying these annuities, been employed during this time, in paying off so much of the debt at *par* every year; extinguishing at the same time every year an equivalent tax, 45 millions would have been paid. But had the savings, also, instead of being sunk as they arose, been employed in the same manner, 71 millions would have been paid.

The nation, therefore, must, without doubt, lose greatly by all schemes of this kind; and yet they have been often much talked of; and, indeed, I shall not wonder, should I hereafter see an attempt made to pay off the national debt in this way.

I must beg leave to detain the reader here some time longer. A more particular explanation of this subject, will lead to some observations on the best methods of raising

---

(*a*) A demonstration of this will be given in the Appendix, note (K).

money which, I think, deserve to be carefully considered.

When any sum is said to be the *value* of a life-annuity, the meaning is, that, in consequence of being improved at interest, and allowing for the chances of mortality, it will bear the whole expence of the annuity. If, therefore, instead of being *laid* up for improvement, it is either immediately applied to particular uses, or has been long since spent; there will be a loss, equal to the sum which would have been added to the purchase-money, had it been *improved*.—This is the reason of the loss which, I have shewn, the public would suffer by offering *life-annuities*, in lieu of *stock*, in order to extinguish its debts. And for the same reason, it must always lose considerably by *raising* money on life-annuities.

Suppose a million raised by *annuities* on a set of lives, all at 30 years of age. Persons at this age have, (according to Tables III, IV, and V,) an *expectation* of 28 years. That is; the duration of their lives, taking them one with another, will be 28 years; (see the beginning of the first Essay.) and they will be entitled, supposing interest at 4 *per cent.* to 7 *l. per annum*, for every 100 *l.* advanced. For a million then, the public would make 28 payments of 70,000 *l.*—Let us suppose next, that a fund producing this sum annually, instead of of being engaged to pay thefe life-annuities, is engaged for 28 years, to pay the principal and intereſt of a million, borrowed on *redeemable* perpetuities, at 4 *per cent*. There will, at the end of the firſt year, be a ſurplus of 30,000 *l*.—In conſequence of applying this to the extinction of the principal, it will be reduced to 970,000 *l*. on which, at the end of the ſecond year, the intereſt due will be 38,800 *l*. There will, therefore, be a ſaving of 1200 *l*. Inſtead of employing this ſaving in further ſinking the *principal*, which would cauſe the fund to accumulate in the ſame manner with money at compound intereſt, let it be taken and employed in any other way: And let the ſame be done with all the ſubſequent ſavings, reſerving only 30,000 *l*. annually, for the purpoſe of ſinking the principal. At the end of the ſecond year, the principal will be 940,000 *l*.; and the ſaving of intereſt upon it, at the end of the third year, 2400 *l*. At the end of the 28th year, the principal will be reduced to 160,000 *l*. The ſaving of intereſt that year will be, 1200 *l*. multiplied by 27, or 32,400; and the ſum of all the ſavings will be 453,600 *l*.—Deduct from hence 160,000 *l*. remaining then undiſcharged of the principal; and 293,600 *l*. will be the loſs the public would ſuſtain, in the circumſtances I have ſuppoſed, by raiſing money on life-annuities. But if we ſuppoſe the ſavings, as they ariſe, as well as the conſtant ſum of 30,000 *l*. to be applied to the discharge of the principal, instead of being spent on current services; the whole million will be annihilated in 21 years and a half; and the loss to the public by life-annuities, will be 6½ years purchase of the annuities; or 455,000*l.*—By similar deductions it may be easily found, that the loss, in *younger* lives, is greater; in *older* lives less; but never inconsiderable, except in the *oldest* lives.

It appears, therefore, that, in consequence of such a way of raising money, the public must always pay much more in interest than there is any occasion for; and *waste* a sum nearly equal to half the principal borrowed (*a*). This, however, tho' so wasteful, is a more frugal

---

(*a*) It is obvious, that the observations here made, may be applied to the common methods of raising money on life-annuities, for building churches, paving streets, making navigations, &c. &c. And, in general, to all cases where the money received, is not laid up to be improved.—For, to view this subject in another light, let us suppose 10,000*l.* borrowed for any public work, on perpetuities, at 4 *per cent.* And, if that will afford more encouragement, let them be made irredeemable for any number of years less than seventeen. Let us further suppose, such rates, or tolls, established for the payment of the interest and principal, as shall produce *double* the interest of the sum borrowed; or 800*l. per annum*, instead of 400 *l. per annum.* Let the *surplus*, as it comes in half-yearly, be laid up to accumulate in the public funds. In 17 years and a half, reckoning interest at 4 *per cent.* a capital will be raised, equal to the whole sum borrowed; and, therefore, at the end of that time, the whole debt may be discharged, and the whole transaction finished.— But if the same sum had been borrowed on annuities, for

the

frugal way of procuring money than by borrowing on *perpetuities*, without putting them into a courſe of redemption; for in this caſe, (if a ſpunge is not applied) the loſs muſt be *infinite*.

I muſt add, that theſe obſervations are particularly applicable to all the ways of raiſing money by the ſale of reverſions.—The public, for inſtance, might procure a million, by offering for it a fund, that will be diſengaged at the end of 18 years; and then produce 80,000 *l. per annum* for ever. This, ſuppoſing intereſt at 4 *per cent*, would be the very ſame with offering *two* millions, 18 years hence, for *one* million now: And a private man, or an *office* for the ſale of reverſions, might gain by ſuch a tranſaction; becauſe, the money advanced, in conſequence of being improved, might, in 18 years, be more than doubled,

the lives of a ſet of perſons 50 years of age, at 8 *per cent*. which is 1 *l. per cent*. leſs than the true value of ſuch annuities: Had this, I ſay, been done, *half* the annuitants would have been alive at the end of the term I have mentioned; (ſee Tables III, IV, and V,) and the whole tranſaction, together with the expences and trouble attending the management of it, could not have been finally cloſed 'till the extinction of all the lives; that is, not in leſs time, moſt probably, than 35, or, perhaps, 40 years.—It is a neceſſary obſervation here, that, if public credit maintains its ground, much will not depend, in the plan now propoſed, on the riſe and fall of STOCKS. If a *war* ſinks them, the money laid out, while the war laſts, will accumulate faſter. If a peace raiſes them, the money that had been previouſly laid out will be proportionably increaſed.

doubled. But, as the *public* always borrows for immediate services, and never lays up money, it would necessarily lose a sum equal to the whole sum borrowed: And the same money might have been borrowed on a fund, producing 50,000*l. per annum*; which would not only pay the interest, but discharge the whole principal in 41 years (*a*).

By raising money on life-annuities, the *present members* of a state take a heavier load on themselves, in order to exempt *posterity*; and there would be a laudable generosity in this, were it not for the *folly* of it; the same exemption being equally practicable at *half* the expence,—On the other hand. By borrowing on *reversionary* grants, the present members of a state exempt themselves *entirely*, by throwing the load *doubled* on posterity; and there is a cruelty and injustice in this that nothing can excuse.

It is well known, that both these methods of raising money have been practised among us. This, however, is, by no means, the worst that has been done. It has been common to borrow money to pay the interest of money borrowed, and thus to give *compound interest* for money; and our *parliaments* have,

---

(*a*) The smallness of the sums, which I have here and elsewhere sometimes supposed to be employed in discharging the public debts, can create no difficulties, because there is no sum which may not be applied to this use by purchasing stock.

some-

sometimes, expressly provided, that this shall be done for a succession of years.

But to return.

The enquiry which has occasioned this digression, must be highly interesting to every person who wishes well to his country.—All schemes for discharging the public debts, by life-annuities, have been shewn to be absurd and extravagant.—In general; it may be observed, that it is far from probable, that any money which the nation can spare, if applied so as to bear only *simple* interest, can be capable of reducing its debts within due bounds; or of doing us, in our present circumstances, any essential service. A fund, producing a surplus of even two millions annually, would, when thus applied, pay no more than 40 millions in 20 years; and, in that time, a war might probably come, which would interrupt the application of it; and increase our debts much more than such a fund had lessened them.

Certain it is, therefore, that if our affairs are to be retrieved, it must be by a *fund* increasing itself in the manner I have explained. The smallest *fund* of this kind is, indeed, *omnipotent*, if it is allowed time to operate. But we are, I fear, got so near to the limits of the resources of the nation, that it cannot be allowed much time: And, in order to make amends for this, it is necessary

that

that it should be *large*.—Let us then suppose, that the nation is still strong enough to enable it to provide a fund, that shall yield a *million* and *half annually*, for 20 years to come: And also, that, together with all its *present* burdens, it is capable of bearing every *additional* burden that 20 years more can bring upon it. If this is not true, we have, I think, nothing to do, but to wait the issue, and tremble.

A fund, producing annually a million and a half, would increase to three millions *per ann.* in 20 years (*a*). At the end of this term, the nation might be eased of the most oppressive taxes, to the amount of a million and a half; and the consequence would prove, that, if there should have been a war, either the whole, or much the greatest part of the addition occasioned by it to the public burdens, would be taken off, and the nation reinstated nearly in its present circumstances. But, if there should have been no war, the national debt, and the taxes charged with it, would be reduced a third below the sums at which they now stand; and the nation would be so much relieved as to be prepared for a war.—The remaining million and half would,

---

(*a*) It should be remembered, that in the year 1781, 1 *l. per cent.* on the consolidated 4 *per cents.* will be annihilated; and that I suppose the savings derived from hence to be taken at that time as a part of the fund.—Methods might be easily contrived for getting this saving immediately, which would be some advantage.

in 23 years, increase again to three millions *per annum*; and then, so much more of the public taxes would be set free; 50 millions more, or 93 millions in all, of the public debts would be discharged, and the difficulties of the nation would be, in a great measure, conquered.—During this whole course of time, there may possibly be but one war; and should that happen, the appropriation at the end of it, of about 400,000 *l. per annum*, might be enough to answer all purposes.

In these observations, I suppose the 3 *per cents*. to be paid off at *par*; and no advantage taken at any time of their low price. By taking this advantage, and with the help of a little management, a fund, producing annually a million and half, might be made to increase to another million and half, in less time than I have assigned. Should there be a war in a few years, the 3 *per cents*. would probably fall below 75; and then the proprietors of them must be glad to part with them at this price; the consequence of which, supposing the war to last eight years, would be, that the fund would double itself, and the nation be relieved in the manner I have mentioned, in 18, instead of 20 years.— The advantage will be the same, supposing the government at such a time to go on in paying off the 3 *per cents* at *par*. For the effect of this would be, that money might be borrowed for the public service on proportionably

ably better terms. Suppose, for instance, that four millions must be borrowed for the service of the year; and let the produce of the fund be then increased to two millions; and the interest of money in the *stocks*, above 4 *per cent*. In these circumstances, it would be the interest of the lenders of money, to take 3¼ *per cent.* for the sums they advanced, in consideration of having their 3 *per cents.* paid off at *par*, to the amount of half these sums.—War, therefore, would accelerate the redemption of the public debts; and it would do this the more, the longer it lasted, and the higher it raised the interest of money: Or if, in consequence of paying always at *par*, this could not happen; an equivalent effect would be produced in the way just mentioned. The stocks would be always kept up by the operations of the fund; and, in proportion to the sums yielded by it, the public would be able to borrow money more advantageously, and less would be added to its burdens.—This seems to me an observation of particular consequence. It demonstrates, that the invariable application, in *war* as well as *peace*, of the produce of the fund I am supposing, to the payment of the national debts, rather than to any current services, would, independently of its effect in (*a*) redeeming these debts, be attended

(*a*) So true is this, that a war, were we now engaged in it, would only render the *present* time so much the more

ed with great advantages to the public. But this is a subject on which I shall have occasion to say more presently. The more proper for entering into measures for paying the public debts. And the following observations will put this out of doubt.

As it is now become the practice to have recourse to *lotteries* in *peace*, we may be sure, that no year will pass without them in *war*. I would, therefore, propose, that, instead of making use of them in raising the annual supplies in war, they should be then applied as an aid in discharging the public debts.—Suppose the war to last 10 years, and the 3 *per cents*. at 70.—Suppose also, each *lottery* to consist of 750,000*l*. in *tickets*, which, when disposed of to subscribers, will bring in 1,050,000*l*. On these suppositions, the whole *loss* to the public, from applying the lotteries to the payment of the public debts, rather than to the current supplies, will be 1,050,000 *l*. annually, or 10 millions and $\frac{1}{2}$ in all.—The *gain* will be as follows: 750,000*l*. of the produce of the sinking fund, formed into tickets, will be the same with 1,050,000*l*.; and this sum will pay off a million and a half of the 3 *per cents*, every year, or 15 millions in all; and the growing savings arising from these payments, will, at the end of 10 years, have paid, at least, two millions more. The nation, therefore, having paid off 17 millions of its debts, and added to them only 10 millions and $\frac{1}{2}$, will gain six millions and $\frac{1}{2}$. But this will be the smallest part of its gains. All the produce of the *sinking fund*, over and above 750,000*l*. might be charged with the payment of the interest of such new debts as would be necessary to be contracted during the war; and, at the end of it, the nation, with the help of 200,000 *l*. to be disengaged in 1781, by the reduction of the 4 *per cents*, would find itself possessed of a *fund*, producing 1,450,000 *l*. annually; which, faithfully employed, might probably be sufficient to extricate it from all its difficulties.—Besides this; such a scheme would not only *preserve*, but *raise* and *establish* the credit of the public: And he only can be duly sensible of the importance of this, who will consider, what danger there would

The *sinking fund,* in its present state, and, after supplying the deficiencies of the peace establishment, yields, I suppose, a considerable part of the million and a half I have mentioned. An annual lottery might easily raise 200,000*l.* more. But this is a measure which I cannot wish to see carried into execution, unless absolutely necessary. Were the managers of our affairs sufficiently in earnest in this business, I cannot doubt but that such savings might be made in the collection and expenditure of the national revenue, as would cause the sinking fund to yield, for 18 or 20 years to come, the *whole* of this sum, without imposing any new burdens on the public. But, were there, indeed, no way of providing any part of it, but by creating new funds, or imposing new taxes; it *ought* to be done, because it *must* be done, or the nation sink.

The evils and dangers, attending an *exorbitant* public debt in this country, are so great, that they cannot be exaggerated.—Without repeating, what has been so often said, of its increasing the dependance on the crown,

would be in another war, should it continue long, of either *overwhelming* public credit; or of being terrified, by the apprehension of such a calamity, into an ignominious and fatal peace. The establishment, therefore, of some such plan as that now proposed, would, at the beginning of a war, be the most important of all works.

render-

rendering us tributary to foreigners; and raising the price of provisions and labour; and, consequently, checking population, and loading our trade and manufactures; I will only take notice of the following evils which attend it.

First. The execrable practices of the *alley*. These cannot be mentioned in language too strong. They are growing every day; and the national debt, by giving occasion to them, is likely soon (with the aid of annual lotteries) to ruin all honest industry among us, and to turn us into a nation of gamblers.

Secondly. It must check the exertions of the spirit of liberty in the kingdom. The tendency of every government is to despotism; and in this it must end, if the people are not constantly jealous and watchful. Opposition, therefore, and resistance, are often necessary. But they may throw things into confusion, and occasion the ruin of the public funds. The apprehension of this must influence all who have their interest connected with the preservation of the funds, and incline them always to acquiescence and servility.

But further. It exposes us to particular danger from *foreign* as well as *domestic* enemies, by making us fearful of war, and incapable of engaging in it, however necessary, without the hazard of bringing on terrible convulsions, by overwhelming public credit.

All these are evils which must increase with every increase of the national debt; and there is a point at which, when they arrive, the consequences must be fatal (*a*).—I am now writing under a conviction, that I am doing the little in my power to preserve my country from this danger. I have shewn, that an annual supply of a million and a half for 18, or at most 20 years, might probably be made the means of restoring and saving us. This, therefore, is our remedy; and it ought to be applied *immediately*, least it should not be applied time enough.

But to proceed to some further observations.

What has been said, has all along supposed a *sacred* and *inviolable* application of the fund I have described, and of all its earnings, to the purpose of sinking the national debt. The whole effect of it depends on its being allowed to operate, WITHOUT INTERRUPTION, a proper time. But it may be asked, how this can be secured? Or, by what method an object, that must be continually growing more and more tempting, can be

---

(*a*) " Either the nation (Mr. *Hume* says, Essays Vol. II. p. 145,) must destroy public credit; or public " credit will destroy the nation."

defended against invasion and rapine?—I might here mention the superintendency and care of the representatives of the kingdom, the faithful guardians of the state, to whom ministers are responsible for the use they make of the public money. But experience has shewn, that we cannot rely on this security.—The difficulty, therefore, now mentioned, is the very greatest difficulty the nation has to struggle with in the payment of its debts.

The sinking fund was established in the year 1716, or soon after the accession of the present family, at a time when the public debts, tho' not much more than a third of what they are now, were thought to be so considerable as to be alarming and dangerous. It was intended as a SACRED DEPOSIT never to be touched; the law which established it declaring, that it was to be applied to the payment of the principal and interest of such national debts and incumbrances, as had been incurred before the 25th of *December* 1716; *and to no other use, intent or purpose whatever.*—The faith of *parliament,* therefore, as well as the security of the kingdom, seemed to require, that it should be preserved carefully and rigorously from alienation. But, notwithstanding this, it has been *generally* alienated; and the produce of it employed, in helping to defray such current expences as the exigences of the state rendered necessary.

In order to justify this, it has been usual to plead, that when money is wanted, it makes no difference, whether it is taken from hence, or procured by making a new loan. There cannot be a worse sophism than this. The difference between these two methods of procuring money is no less than *infinite*.—For, let us suppose, a *million* wanted for any public service. If it is borrowed at 4 *per cent.* the public will lose by the payment of interest 40,000 *l.* the first year, and the same the second year, and the same for ever afterwards. But if it is taken out of the *sinking fund*, the public will lose 40,000*l.* the first year; 40,160 *l.* the second year; 80,000*l.* the 18th year; a *million* the 85th year: For these are the sums that would, at these times, have otherwise necessarily reverted to the public. It loses, therefore, the advantage of paying in 85 years, with money of which otherwise no use could have been made, *twenty-five* millions of debt.—In other words; by employing the SINKING FUND, in bearing current expences, rather than borrowing *new* money on new funds; the state, in order to avoid giving *simple interest* for money, is made to alienate money, that *must* have otherwise been improved at *compound interest*; and that, in time, would have *necessarily* increased to *any sum*.

sum.—Had a faithful use been made from the first, of only one THIRD of the produce of this fund, the greatest part of our present debts would now have been discharged (*a*).—Can it be possible then to think, without regret and indignation, of that misapplication of this fund, which, with the consent of parliaments always complying, our ministers have practised?—It is difficult here to speak with calmness.—But I forbear.—*Calculation*, and not *censure*, is my business in this work.

(*a*) See the *Questions* at the end of the *Appendix*.
The principal observations in this Chapter, I have given just as they occurred to my thoughts, without knowing that any of them had been made by other writers. Some proposals and observations of a similar nature, I have since found in an excellent pamphlet published in 1726, entitled, *An Essay on the National Debts of this kingdom, wherein the importance of discharging them is considered, and some general mistakes about the nature and efficacy of the* SINKING FUND *examined and removed. In a Letter to a Member of the House of Commons*. Fourth edition.

I must beg leave to add, that in a pamphlet published since the former editions of this Treatise, and entitled, *An Appeal to the Public on the Subject of the National Debt*, I have endeavoured to explain such parts of this chapter as have been thought not sufficiently clear; and given a more full account of the *nature*, *powers* and *history* of the *Sinking Fund*, and of the *pernicious consequences* of those alienations of it which I have censured above, and which for many years, have made a part of the fixed practice of government among us.

ESSAY I.*

Containing Observations on the Expectations of Lives; the Increase of Mankind; the Number of Inhabitants in LONDON; *and the Influence of great Towns, on Health and Population.*

In a LETTER *to* BENJAMIN FRANKLIN, *Esq*; L.L.D. *and* F.R.S.

DEAR SIR,

I Beg leave to submit to your perusal the following observations. If you think them of any importance, I shall be obliged to you for communicating them to the Royal Society. You will find, that the chief subject of them is the present state of the city of *London*, with respect to healthfulness and number of inhabitants, as far as it can be collected from the bills of mortality. This is a subject that has been considered by others; but the proper method of calculating

---

* This Essay was read to the ROYAL SOCIETY, April 27th, 1769, and has been published in the Philosophical Transactions, Vol. 59. It is here republished with corrections; and several additions, particularly the *Postscript*.

from the bills has not, I think, been sufficiently explained.

No competent judgment can be formed of the following observations, without a clear notion of what the writers on *Life-Annuities* and *Reversions* have called the *Expectation of Life*. Perhaps this is not in common properly understood; and Mr. *De Moivre*'s manner of expressing himself about it is very liable to be mistaken.

The most obvious sense of the *expectation* of a given life is, "That particular number of years which a life of a given age has an equal chance of enjoying." This is properly the time that a person may reasonably *expect* to live; for the chances *against* his living longer are greater than those *for* it; and, therefore, he cannot entertain an *expectation* of living longer, consistently with probability. This period does not coincide with what the writers on Annuities call the *expectation of life*, except on the supposition of an uniform decrease in the probabilities of life, as Mr. *Simpson* has observed in his *Select Exercises*, p. 273.—It is necessary to add, that, even on this supposition, it does not coincide with what is called the *expectation of life*, in any case of joint lives. Thus, two lives of 40 have an even chance, according to Mr. *De Moivre*'s hypothesis (*a*), of continuing together only $13\frac{1}{2}$ years. But the *expectation*

(*a*) See the Notes in page 2 and 23.

of two equal joint lives, being (according to the same hypothesis) always a *third* of the *common complement*; it is, in this case, $15\frac{1}{3}$ years. It is necessary, therefore, to observe, that there is another sense of this phrase, which ought to be carefully distinguished from that now mentioned. It may signify, "The *mean continuance* of any given *single, joint*, or *surviving* lives, according to any given Table of Observations:" that is, the number of years which, taking them one with another, they actually enjoy, and may be considered as sure of enjoying; those who live or survive *beyond* that period, enjoying as much *more* time in proportion to their number, as those who *fall short* of it enjoy *less*. Thus; Supposing 46 persons alive, all 40 years of age; and that, according to Mr. *De Moivre's hypothesis*, one will die every year 'till they are all dead in 46 years; half 46, or 23, will be their *expectation of life*: That is; The number of years enjoyed by them all, will be just the same as if every one of them had lived 23 years, and then died; so that, supposing no interest of money, there would be no difference in value between annuities payable for life to every single person in such a set, and equal annuities payable to another equal set of persons of the same common age, supposed to be all sure of living just 23 years and no more.

In

In like manner; the *third* of 46 years, or 15 years and 4 months (a), is the *expectation* of two joint lives both 40; and this is also the *expectation* of the survivor. That is; supposing a set of marriages between persons all 40, they will, one with another, last just this time; and the survivors will last the same time. And annuities payable during the continuance of such marriages would, supposing no interest of money, be of exactly the same value with annuities to begin at the extinction of such marriages, and to be paid, during life, to the survivors.—In adding together the years which any great number of such marriages, and their survivorships, have lasted, the sums would be found to be equal.

One is naturally led to understand the *expectation* of life in the first of the senses now explained, when, by Mr. *Simpson* and Mr. *De Moivre*, it is called, *the number of years which, upon an equality of chance, a person may expect to enjoy*; or, *the time which a person of a given age may justly expect to continue in being*; and, in the last sense, when it is called, *the share of life due to a person*. But, as in reality it is always used in the last of these senses, the former language should not be applied to it: And it is in this last sense, that it coincides with the *sums* of the *present* probabilities, that any given single or joint lives shall attain to the end of the

(a) See Note (L) Appendix.

1st, 2d, 3d, &c. *moments*, from this time to the end of their possible existence; or, (in the case of survivorships) with the sum of the probabilities, that there shall be a survivor at the end of the 1st, 2d, 3d, &c. moments, from the present time to the end of the possible existence of survivorship. This coincidence every one conversant in these subjects must see, upon reflecting, that both these senses give the true present value of a life-annuity, secured by land, without interest of money (*a*).

This period in joint lives, I have observed is never the same with the period which they have an equal chance of enjoying; and in single lives, I have observed, they are the same only on the supposition of an uniform decrease in the probabilities of life. If this decrease, instead of being always uniform, is *accelerated* in the last stages of life; the former period, in single lives, will be *less* than the latter; if *retarded*, it will be *greater*.

It is necessary to add, that the number expressing the former period, multiplied by the number of single or joint lives whose expectation it is, added annually to a society or town, gives the whole number living together, to which such an annual addition would in time grow. Thus; since 19, or the third of 57, is the *expectation* of two

(*a*) See Note (L) in the Appendix.

joint lives, whose common age is 29, or common *complement* 57; twenty marriages every year between persons of this age would, in 57 years, grow to 20 times 19, or 380 marriages always existing together. The number of *survivors* also arising from these marriages, and always living together, would, in twice 57 years, increase to the same number. And, since the *expectation* of a single life is always half its *complement*; in 57 years likewise, 20 single persons aged 29, added annually to a town, would increase to 20 times 28.5 or 570; and, when arrived at this number, the deaths every year will just equal the accessions, and no further increase be possible.

It appears from hence, that the particular proportion that becomes extinct every year, out of the whole number constantly existing together of single or joint lives, must, wherever this number undergoes no variation, be exactly the same with the *expectation* of those lives, at the time when their existence commenced. Thus; was it found that a 19th part of all the marriages among any body of men, whose numbers do not vary, are dissolved every year by the deaths of either the husband or wife, it would appear that 19 was, at the time they were contracted, the *expectation* of these marriages. In like manner; was it found in a society, limited to a fixed number of members,

members, that a 28th part dies annually out of the whole number of members, it would appear that 28 was their common expectation of life at the time they entered. So likewise; were it found in any town or district, where the number of births and burials are equal, that a 20th or 30th part of the inhabitants die annually, it would appear, that 20 or 30 was the *expectation* of a child just born in that town or district. These *expectations*, therefore, for all single lives, are easily found by a *Table of Observations*, shewing the number that die annually at all ages, out of a given number alive at those ages; and the general rule for this purpose, is " to divide the sum of all the living in the Table, at the age whose expectation is required, and at all greater ages, by the sum of all that die annually at that age, and above it; or, which is the same, by the number (in the Table) of the living at that age; and half unity subtracted from the quotient will be the required *expectation* (*a*)." Thus, in Dr. *Halley*'s Table, the sum of all the living at 20 and upwards is, 20,724. The number living at that age is 598; and the former

---

(*a*) This rule, and also rules for finding in all cases the expectations of joint lives and survivorships, may be deduced with great ease, by having recourse to the doctrine of fluxions. In this method, Mr. *De Moivre* says, he discovered them. See Appendix, note (L), where an account will be given of these deductions, omitted by Mr. *De Moivre*.

number

number divided by the latter, and half unity (*a*) subtracted from the quotient, gives 34.15 for the *expectation* of 20. The expectation of the same life by Mr. *Simpson*'s Table, formed from the bills of mortality of *London*, is 28.9 (*b*).

These

(*a*). If we conceive the *recruit* necessary to supply the *waste* of every year to be made always at the *end* of the year, the *dividend* ought to be the *medium* between the numbers living at the *beginning* and the *end* of the year. That is, it ought to be taken *less* than the sum of the living in the Table at and above the given age, by *half* the number that die in the year; the effect of which *diminution* will be the same with the *subtraction* here directed.— The reason of this subtraction will be further explained, in the beginning of the last Essay.

(*b*) It appears in p. 169 and 170, that the *expectations* of *single* and *joint* lives are the same with the values of *annuities* on these lives, supposing no interest or improvement of money.—In considering this subject, it will, probably, occur to some, that, allowing interest for money, the values of lives must be the same with the values of annuities *certain* for a number of years equal to the *expectations* of the lives. But care must be taken not to fall into this mistake. The latter values are always greater than the former: And the reason is, that, tho' a number of *single* or *joint* lives of given ages will, among them, enjoy a *given* number of years, yet some of them will enjoy a much *greater*, and some a much *less* number of years. Thus; 100 marriages among persons, all 29, would, as I have said, one with another, exist 19 years; and an office bound to pay annuities to such marriages during their continuance, might reckon upon making 19 payments for each marriage. But then, many of these payments would not be made 'till the end of 30, and some not 'till the end of 40 years. And it is apparent, that on account of the greater value of *quick* than *late* payments, when money *bears* interest, 19 payments so made cannot be

worth

These observations bring me to the principal point which I have had all along in view. They suggest to us an easy method of finding the number of inhabitants in a place, from a *Table of Observations*, or the *bills of mortality* for that place, supposing the yearly births and burials equal. " Find by
" the Table, in the way just described, the
" *expectation* of an infant just born, and this,
" multiplied by the number of yearly births,
" will be the number of inhabitants." At *Breslaw*, according to Dr. *Halley*'s Table, though half die under 16, and therefore an

worth as much, as the same number of payments made regularly at the end of every year, 'till in 19 years they are all made.

This observation might be employed, to demonstrate further, the error of those who have maintained, that the value of a given life is the same, with the value of an annuity certain, for as many years as the life has an equal chance of existing. Were this true, an annuity on a life, supposed to be exposed to such danger in a particular year, as to create an equal chance, whether it will not fail that year, would, at the beginning of the year, be worth *nothing*, though supposed to be sure of continuing for ever, if it escaped that danger: nor, in general, would the values of annuities on a set of lives, be at all affected by any alterations in the rate of mortality among them, provided these alterations were such, as did not affect the period during which they had an equal chance of existing. —But there can be no occasion for taking notice of an opinion, which has been embraced only by persons ignorant of mathematics, and plainly unacquainted with the genuine principles of calculation on this subject.— See a pamphlet on Life-Annuities by *Weyman Lee*, Esq; of the *Inner Temple*.

infant juft born has an *equal chance* of living only 16 years; yet his *expectation*, found by the rule I have given, is near 28 years; and this, multiplied by 1238, the number born annually, gives 34,664, the number of inhabitants. In like manner, it appears from Mr. *Simpfon*'s Table, that, though an infant juft born in *London* has not an *equal chance* of living 3 years, his *expectation* is 20 years; and this number, multiplied by the yearly births, would give the number of inhabitants in *London*, were the births and burials equal.— The medium of the yearly births, for ten years, from 1759 to 1768, was 15,710. This number multiplied by 20, is 314,200; which is the number of inhabitants that there would be in *London*, according to the bills, were the yearly burials no more than equal to the births: that is, were it to fupport itfelf in its number of inhabitants, without any fupply from the country. But for the period I have mentioned, the burials were, at an average, 22,956, and exceeded the chriftenings 7,246. This is, therefore, at prefent, the yearly addition of people to *London* from other parts of the kingdom, by whom it is kept up. Suppofe them to be all, one with another, perfons who have, when they remove to *London*, an *expectation* of life equal to 30 years. That is; fuppofe them to be all of the age of 18 or 20, a fuppofition certainly far beyond the truth. From hence will arife, according

cording to what has been before obferved, an addition of 30, multiplied by 7.246; that is, 217,380 inhabitants. This number, added to the former, makes 531,580; and this, I think, at moft, would be the number of inhabitants in *London* were the bills perfect. But it is certain, that they give the number of births and burials too little. There are many burying places that are never brought into the bills. Many alfo emigrate to the navy and army and country; and thefe ought to be added to the number of deaths. What the deficiencies arifing from hence are, cannot be determined. Suppofe them equivalent to 6000 every year in the births, and 6000 in the burials. This would make an addition of 20 times 6000, or 120,000, to the laft number; and the whole number of inhabitants would be 651,580. If the burials are deficient only two-thirds of this number, or 4000; and the births the whole of it; 20 multiplied by 6000, muft be added to 314,290, on account of the defects in the births: And, fince the excefs of the burials above the births will then be only 5,246; 30 multiplied by 5,246 or 157,380, will be the number to be added on this account; and the fum, or number of inhabitants, will be 591,580.—But if, on the contrary, the burials are deficient 6000, and the births only 4000; 80,000 muft be added to 314,290, on account of the deficiencies in the births;

and 30 multiplied by 9,246, or 277,380, on account of the excefs of the burials above the births; and the whole number of inhabitants will be 671,580.

Every fuppofition in thefe calculations is too high. *Emigrants* from *London* are, in particular, allowed the fame *expectation* of continuance in *London* with thofe who are born in it, or who come to it in the firmeft part of life, and never afterwards leave it; whereas, it is not credible that the former *expectation* fhould be fo much as half the latter. But I have a further reafon for thinking that this calculation gives too high numbers, which has with me irrefiftible weight. It has been feen, that the number of inhabitants comes out lefs on the fuppofition, that the defects in the chriftenings are greater than thofe in the burials. Now it feems evident that this is really the cafe; and, as it is a fact not attended to, I will here endeavour to explain diftinctly the reafon which proves it.

The proportion of the number of births in *London*, to the number who live to be 10 years of age, is, by the bills, 16 to 5. Any one may find this to be true, by fubtracting the *annual medium* of thofe who have died under 10, for fome years paft, from the *annual medium* of births for the fame number of years.—Now, tho' without doubt, *London* is very fatal to children, yet it feems incredible

credible that it should be so fatal as this implies. The *bills*, therefore, probably, give the number of those who die under 10 too great in proportion to the number of births; and there can be no other cause of this, than a greater deficiency in the *births* than in the *burials*. Were the deficiencies in both equal; that is, were the *burials*, in proportion to their number, just as deficient as the *births* are in proportion to *their* number, the proportion of those who reach 10 years of age to the number born, would be right in the *bills*, let the deficiencies themselves be ever so considerable. On the contrary; were the deficiencies in the *burials* greater than in the *births*, this proportion would be given too great; and it is only when the former are least, that this proportion can be given too little.—Thus; let the number of annual *burials* be 23,000; of *births* 15,700; and the number dying annually under 10, 10,800. Then 4,900 will reach 10, of 15,700 born annually; that is, 5 out of 16. —Were there no deficiencies in the *burials*, and were it fact that only *half* the number born die under 10; it would follow, that there was an annual deficiency equal to 4,900 subtracted from 10,800, or 5,900, in the *births*.—Were the *births* a third part too little, and the *burials* also a third part too little, the true number of *births*, *burials*, and of *children dying under* 10, would be 20,933--30,666 and

and 14,400; and, therefore, the number that would live to 10 years of age, would be 6,533 out of 20,933, or 5 of 16 as before.—Were the *births* a third part, and the *burials* so much as two-fifths wrong, the number of *births*, *burials*, and children dying under 10 would be 20,933—32,200—and 15,120. And, therefore, the number that would live to 10 would be 5,813 out of 20,933, or five out of 18.—Were the *births* a third part wrong, and the *burials* but a 6th, the foregoing numbers would be 20,933—26,833—12,600; and therefore, the number that would live to 10 would be 8,333 out of 20,933, or 5 out of 12.56: and this proportion seems as low as is consistent with probability. It is somewhat less than the proportion in Mr. *Simpson*'s Table of *London Observations*; and much less than the proportion in the Table of *Observations* for *Breslaw*. The deficiencies, therefore, in the register of *births*, must be greater than those in the register of *burials* (a); and the least num-

(a) One obvious reason of this fact is, that *none* of the *births* among *Jews*, *Quakers*, *Papists*, and the *three denominations of Dissenters* are included in the bills, whereas *many* of their *burials* are. It is further to be attended to, that the abortive and still-born, amounting to about 600 annually, are included in the burials, but never in the births. If we add these to the christenings, preserving the burials the same, the proportion of the born according to the bills, who have reached ten for the last sixteen years, will be very nearly one *third* instead of *five sixteenths*.

ber I have given, or 591,580 is nearest to the true number of inhabitants. However, should any one, after all, think that it is not improbable that only 5 of 16 should live in *London* to be 10 years of age; or that above *two-thirds* die under this age; the consequence will still be, that the foregoing calculation has been carried too high. For it will from hence follow, that the *expectation* of a child just born in *London* cannot be so much as I have taken it. This *expectation* is 20, on the supposition that half die under 3 years of age, and that 5 of 16 live to be 29 years of age, agreeably to Mr. *Simpson*'s Table. But if it is indeed true, that *half* die under 2 years of age, and 5 of 16 under 10, agreeably to the *bills*, this expectation cannot be so much as 17 (a); and all the numbers before given will be considerably reduced.

Upon the whole: I am forced to conclude from these observations, that the second number I have given, or 651,580, though short of the number of inhabitants commonly supposed in *London*, is, very probably, much *greater*, but cannot be *less*, than the true number. Indeed, it is in general evident, that in cases of this kind numbers are very much over-rated. The inge-

---

(a) This may be deduced from the observations in the last Essay; and it will be there proved, that, in reality, this expectation does not exceed 18.

nious Dr. *Brakenridge*, 14 years ago, when the bills were lower than they are now, from the number of houses, and allowing six to a house, made the number of inhabitants 751,800. But his method of determining the (*a*) number of houses is too precarious; and, besides, 6 to a house is probably too large an allowance.—Many families now have two houses to live in.—The magistrates of *Norwich*, in 1752, took an exact account of both the number of houses and indivi-

---

(*a*) Vid. Phil. Transactions, Vol. XLVIII, p. 788. In a paper subsequent to this, Dr. *Brakenridge* tells us, that in a late survey it appeared, that in all *Middlesex*, *London*, *Westminster*, and *Southwark*, there were 87,614 houses, of which 19,324 were cottages, and 4810 empty. And he acknowledges, that this, if right, proves *London* to be much less populous than he had made it. See Phil. Transf. Vol. L, p. 471. He does not mention how this survey was taken; but most probably it must have been incorrect.—Mr. *Maitland* gives two accounts of the number of houses within the bills. One carefully taken from the books of all the parishes and precincts belonging to *London*; and another taken from a particular survey in 1737, made by himself with incredible pains. The first account makes the number of houses 85,805. The second account makes it 95,968. And the reason of the difference he observes, is, that many landlords of small places, paying all taxes, they are in the parish books reckoned as so many single houses, though each of them contain several houses. See Mr. *Maitland*'s History of *London*, 2d Book at the end.—This, perhaps, may be also the reason of the deficiencies which, I suppose, there must be in the survey, mentioned by Dr. *Brakenridge*.—It will be observed presently, that the number of inhabitants in *London* in 1737, was considerably greater than it is now.

duals

duals in that city. (a) The number of houses was 7,139, and of individuals 36,169, which gives nearly 5 to a house.———Another

(a) Vid. Gentleman's Magazine for 1752, and Dr. *Short's Comparative History of the Increase of Mankind*, p. 38. In page 58 of this last work the author says, that, in order to be fully satisfied about the number of persons to be allowed to a family, he procured the true number of families and individuals in 14 market towns, some of them considerable for trade and populousness; and that in them were 20,371 families, and 97,611 individuals, or but little more than $4\frac{3}{4}$ to a family. He adds, that, in order to find the difference in this respect between towns of trade and country parishes, he procured, from divers parts of the kingdom, the exact number of *families* and *individuals*, in 65 country parishes. The number of *families* was 17,208; *individuals* 76,284; or not quite $4\frac{1}{2}$ to a family.—In the place I have just referred to, in the Gentleman's Magazine, there is an account of the number of *houses* and *inhabitants* in *Oxford*, exclusive of the colleges; and in *Wolverhampton, Coventry* and *Birmingham*, for 1750. The number of persons to a *house* was, by this account, $4\frac{2}{3}$ in the two former towns, and $5\frac{1}{4}$ in the two latter.—Dr. *Davenant*, from Mr. *King's* Observations, gives $4\frac{1}{13}$, as the number of persons to a *family* for the whole kingdom. See *An Essay on the probable Method of making a people gainers by the balance of trade.*—The number of *families* in *Rome* in 1740, was 32,158; of *inhabitants* 140,080; or $4\frac{1}{2}$ to a *family*.—In 59 Dutch villages, mentioned by *Struyk*, the number of houses was 12,005; of inhabitants, 45,888, or not 4 to a *house*. See *Susmilch's Gottliche Ordnung*, or a Treatise in German on the Probabilities of Human Life in different situations, population, &c. Vol. I. p. 233.—In the whole province of VAUD in *Switzerland*, the number of persons to a family is $4\frac{1}{2}$. See the beginning of the *Supplement*.—From an account taken in 1770, it appeared, that the number of inhabitants at *Leeds* in *Yorkshire*, was 16,380, and of families 3,899. In this populous and opulent town,

ther method which Dr. *Brakenridge* took to determine the number of inhabitants in *London*

therefore, the number of persons in a *family*, is only $4\frac{1}{3}$: And the number in each *house*, will not be quite 5, supposing every *fifth house* to contain *two families*.—From an account with which a friend at *Shrewsbury* has favoured me, it appears, that in that town, in 1750, the number of inhabitants to a *house* was $4\frac{1}{3}$.—Very exact accounts, of which I shall take further notice, prove, that in the parish of *Holy-Cross*, one of the suburbs of *Shrewsbury*, and at *Northampton*, the same proportion is $4\frac{1}{3}$ to a *house* in the former; and $4\frac{3}{4}$ in the latter.—In the parish of *Ackworth* in *Yorkshire*, the number of inhabitants of all ages, in 1757, was 603. In 1767, this number was 728. The number of *houses* in the former year was 160; in the latter year, 184. In the town of *Newbury* in *Berkshire*, the number of inhabitants, according to an account taken in 1768, was 3732; and the number of *houses* 930. In the parish of *Speen*, adjoining to *Newbury*, the number of *inhabitants* in 1757, was 1260; of *houses*, 303. There are, therefore, in each of these three last places, only *four* inhabitants to a *house*.—In the parish of *Aldwinckle*, *Northamptonshire*, the number of *houses* is 96, of *inhabitants* 402; or $4\frac{1}{3}$ to a *house*.—In 1757, the inhabitants of *Manchester* were numbered, and found to be 19,839. They have since increased near 3000; and the number of *houses* is now, I am informed, 4860. In this town, therefore, the number of *inhabitants* to a *house* cannot be above $4\frac{3}{4}$. The same appears to be true of *Liverpool*.——It seems, therefore, that five persons to a house may not be much too small an allowance for *London*, but is too large for *England* in general. From whence it will follow, that Dr. *Brakenridge* has likewise over-rated the number of people in *England*. In a letter to *George Lewis Scott*, Esq; published in 1756, in the Phil. Transf. Vol. 49, p. 877, he says, that he had been certainly informed, that the number of houses rated to the window-tax was 690,000. The number of cottages not rated, he adds, was not accurately known; but from the

accounts

don was from the annual number of burials, adding 2000 to the bills for omiſſions, and ſuppoſing a 30th part to die every year. In order to prove this to be a moderate ſuppoſition he obſerves that, according to Dr. *Halley*'s Obſervations, a 34th part die every year at *Breſlaw*. But this obſervation was made too inadvertently. The number of annual burials there, according to Dr. *Halley*'s account, was 1174, and the number of inhabitants, as deduced by him from his Table, was 34,000; and therefore a 29th part died every year. Beſides; any one may find, that in reality the Table is conſtructed on the ſuppoſition, that the whole

accounts given in, it appeared, that they could not exceed 200,000; and from theſe data, in conſequence of allowing ſix to a houſe, he makes the number of people in *England* to be 5,340,000. Dr. *Brakenridge* has here under-rated the cottages; and the true number of houſes in the kingdom in 1766, was 980,692. See the latter end of the firſt part of the *Supplement*. Call them, however, a million, and the number of people in *England* and *Wales* will be four millions and a half, allowing $4\frac{1}{2}$ to a houſe; and 5 millions, allowing 5 to a *houſe*.—The former is *probably* too large an allowance; but the latter is *certainly* ſo. The number of people in *England* may, therefore, be ſtated as *probably* not more than 4 millions and a half; but *certainly* not 5 millions.—The number of *houſes* in *Ireland* in 1754, was 395,439. In 1767, it was 424,046. (See the *Gentleman's and Citizen's Almanack* for 1772, by *Samuel Watſon, Dublin*). Let $4\frac{1}{2}$ be allowed to a houſe, and the number of people in *Ireland* will be 1,908,207. And, if a million and a half are ſuppoſed in *Scotland*, the number of people in *Great Britain* and *Ireland* will be about eight millions.

number born, or 1238, die every year; from whence it will follow that a 28th part died every year (*a*). Dr. *Brakenridge*, therefore, had he attended to this, would have stated a 24th part as the proportion that dies in *London* every year, and this would have taken off 150,000 from the number he has given. But even this must be less than the just proportion. For let three-fourths of all who either die in *London* or migrate from it, be such as have been born in *London*; and let the rest be persons who have removed to *London* from the country, or from foreign nations. The *expectation* of the former, it has been shewn, cannot exceed 20 years; and 30 years have been allowed to the latter. One with another, then, they will have an *expectation* of $22\tfrac{1}{2}$ years. That is; one of $22\tfrac{1}{2}$ will die every year (*b*). And, consequently,

---

(*a*) Care should be taken, in considering Dr. *Halley*'s Table, not to take the first number in it, or 1000, for so many just born. 1238, he tells us, was the annual medium of births, and 1000 is the number he supposes all living at one year and under. It was inattention to this that led Dr. *Brakenridge* to his mistake.

It will be shewn in the 4th Essay, that the number of the living, under 20, is given too high in this Table; and from hence it will follow, that more than a 28th part of the inhabitants die at *Breslaw* annually.

(*b*) The whole number of inhabitants in *Rome* in 1743, was 147,476, and the annual medium of burials for three years, from 1741 to 1743; was 6338. A 23d part, therefore,

quently, supposing the annual recruit from the country to be 7000, the number of births

fore, died every year. See *Susmilch's Gottliche Ordnung*, quoted p. 183.

In 1761, the whole number of inhabitants in the same town, was 157,452. The annual medium of births for three years, from 1759 to 1761, was 5167; and of burials 7153. One in 22, therefore, died annually. See Dr. *Short's Comparative History of the Increase and Decrease of Mankind in* England *and several Countries abroad*, p. 59, 60.—In 1752, the accurate and diligent Mr. *Struyk*, took particular pains to determine the number of inhabitants in *Amsterdam*; and the result of his enquiry was, that very probably it did not amount to 200,000. The annual medium of burials for six years, from 1747 to 1752, was 8247. *One* in 24, therefore, died annually. See *Susmilch*, ibid.—At *Amsterdam*, there is a great number of Jews, and their burials are not included in the bills. There must, I suppose, be other deficiencies, and an allowance for these would, I doubt not, increase the proportion of inhabitants who die annually, to one in 21 or 22.—At *Dublin*, in the year 1695, the number of inhabitants was found, by an exact survey, to be 40,508, (See Philos. Transactions, No. 261). I find no account of the annual burials just at that time; but from 1661 to 1681, the medium had been 1613; and from 1715 to 1728 it was 2123. There can, therefore, be no material error in supposing that, in 1695, it was 1800; and this makes 1 in 22 to die annually. See Dr. *Short's Comparative History*, p. 15, and *New Observations*, p. 228. —The annual medium of burials for five years, from 1755 to 1759, in *Manchester* and *Salford*, exclusive of those among Dissenters, was 743; of *births*, 756. The number of inhabitants in 1757 was 19,839. See Note, p. 184. Of these *at least* 1500 or 2000 were Dissenters.. About a 24th part, therefore, died annually. But it should be considered here, that *Manchester* has increased so fast by accessions from the country, as to have more than doubled

births 3 times 7000 or 21,000, and the *burials* and *migrations* 28,000 (which are all high

ed itself since 1717; and that the effect of such an increase must be to raise the proportion of *inhabitants* to the *deaths*, and also the proportion of the *births* and *weddings* to the *burials*, higher than they would otherwise be.—The annual medium of burials in the parish church and chapels of LEEDS, from 1754 to 1768, was 758. The number of inhabitants is 16380. See Note, p. 183. One in 21¾ of the inhabitants, therefore, die annually.—These facts prove that I have been too moderate in making only 1 in 22½, including emigrants, to die in *London* annually.

In 1631 the number of people in the *city and liberties* of *London* was taken, by order of the Privy Council, and found to be 130,178.—This account was taken five years after a plague that had swept off near a quarter of the inhabitants; and when, therefore, the town being full of recruits in the vigour of life, the medium of annual burials must have been lower than usual, and the births higher. Could, therefore, the medium of annual burials at that time, within the walls, and in the 16 parishes without the walls, be settled, exclusive of those who died in such parts of the 16 parishes without the walls, as are not in the *liberties*, the proportion dying annually obtained from hence might be depended on, as less than the common and just proportion. But this medium cannot be discovered with any accuracy. *Graunt* estimates that two-thirds of these 16 parishes are within the *liberties*; and, if this is right, the medium of annual burials in the *city and liberties* in 1631, was 5,500, and 1 in 23¾ died annually; or making a small allowance for deficiencies in the bills, 1 in 22.—Mr. *Maitland*, in his History of *London*, Vol. II. page 744, by a laborious, but too unsatisfactory, investigation, reduces this proportion to 1 in 24½; and on the suppositions, that this is the true proportion dying annually, *at all times*, in *London*, and that the deficiencies in the burials (including the burials in *Marybone* and *Pancrafs* parishes) amount

high fuppofitions), the number of inhabitants will be, $22\frac{1}{4}$ multiplied by 28,000, or 630,000.

I will juft mention here one other inftance of exaggeration on the prefent fubject.

Mr. *Corbyn Morris*, in his ufeful *Obfervations on the paft growth and prefent ftate of the city of London*, publifhed in 1751, fuppofes that no more than a 60th part of the inhabitants of *London*, who are above 20, die every year, and from hence he concludes that the number of inhabitants was near a million. In this fuppofition there was an error of at leaft one half. According to Dr. *Halley*'s Table, it has been fhewn, that a 34th part of all at 20 and upwards, die every year at *Breflaw*. In *London*, a 29th part, according to Mr. *Simpfon*'s Table, and alfo according to all other Tables of *London* Obfervations. And in *Scotland* it has been found for many years, that, of 974 minifters and profeffors whofe

to 3,038 annually; he determines, that the number of inhabitants within the bills was 725,903, in the year 1737.

The number of burials not brought to account in the bills is, probably, now much greater than either Dr. *Brakenridge* or Mr. *Maitland* fuppofe it. I have reckoned it fo high as 6000, in order to include emigrants, and alfo to be more fure of not falling below the truth.

It will appear in the laft Effay, with an evidence little fhort of demonftration, that, at leaft, 1 in $20\frac{3}{4}$ die annually in *London*, and that, confequently, the number of inhabitants, if the omiffions in the burials are 6000, cannot exceed 601,750.

ages

ages are 27 and upwards, a 33d part have died every year. Had, therefore, Mr. *Morris* stated a 30th part of all above 20 dying annually in *London*, he would have gone beyond the truth, and his conclusion would have been 400,000 less than it is.

Dr. *Brakenridge* observed, that the number of inhabitants, at the time he calculated, was 127,000 less than it had been. The bills have lately advanced a little, but still they are much below what they were from 1717 to 1743. The medium of the annual *births*, for 20 years, from 1716 to 1736, was 18,000, and of *burials* 26,529; and, by calculating from hence on all the same suppositions with those which made 651,580 to be the present number of inhabitants in *London*, it will be found that the number then was 735,840, or 84,260 greater than the number at present. *London, therefore*, for the last 30 years, has been decreasing; and though now it is increasing again, yet there is reason to think that the additions lately made to the number of buildings round it, are owing, chiefly to the increase of luxury, and the inhabitants requiring more room to live upon (*a*). It

---

(*a*) The medium of annual burials in the 97 parishes within the walls was,

From 1655 to 1664,	3264
From 1680 to 1690,	3139
From 1730 to 1740,	2316
From 1758 to 1768,	1620

This

It should be remembered, that the number of inhabitants in *London* is now so much less as I have made it, than it was 40 years ago, on the supposition, that the proportion of the omissions in the *births* to those in the *burials*, was the same then that it is now. But it appears that this is not the fact.—From 1728, (the year when the ages of the dead were first given in the *bills*) to 1742, near five-sixths of those who were born died under 10, according to the bills. From 1742 to 1752 three quarters: And ever since 1752, this proportion has stood nearly as it is now, or at somewhat more than two-thirds. The omissions in the *births*, therefore, compared with those in the *burials*, were greater formerly; and this must render the difference between the number of inhabitants now and

This account proves, that though, since 1655, *London* has doubled its inhabitants, yet, *within the walls*, they have decreased; and so rapidly for the last 30 years as to be now reduced to one half.—The like may be observed of the 17 parishes immediately without the walls. Since 1730, these parishes have been decreasing so fast, that the *annual burials* in them have sunk from 8,672 to 5,432, and are now lower than they were before the year 1660. In *Westminster*, on the contrary, and the 23 out-parishes. In *Middlesex* and *Surrey*, the *annual burials* have since 1660 advanced from about 4000 to 16,000.——These facts prove, that the inhabitants of *London* are now much less crowded together than they were. It appears, in particular, that *within the walls* the inhabitants take as much room to live upon as double their number did formerly. —The very same conclusions may be drawn from an examination of the *christenings*.

formerly

formerly somewhat less considerable than it may seem to be from the face of the bills. One reason, why the proportion of the amounts of the *births* and *burials* in the bills, comes now nearer than it did, to the true proportion, may, perhaps, be, that the number of Dissenters is lessened. The Foundling Hospital also may have contributed a little to this event, by lessening the number given in the bills as having died under 10, without taking off any from the *births*; for all that die in this hospital are buried at *Pancrafs* church, which is not within the *bills*. See the preface to a collection of the yearly bills of mortality from 1657 to 1758 inclusive, p. 15.

I will add, that it is probable that *London* is now become less fatal to children than it was; and that this is a further circumstance which must reduce the difference I have mentioned; and which is likewise necessary to be joined to the greater deficiencies in the births, in order to account for the very small proportion of children who survived 10 years of age, during the two first of the periods I have specified.—Since 1752, *London* has been thrown more open. The custom of keeping country-houses, and of sending children to be nursed in the country, has prevailed more. But, particularly, the destructive use of spirituous liquors among the poor has been checked.

I have

I have shewn that in *London*, even in its present state, and according to the most moderate computation, half the number born die under *three* years of age. In *Vienna* under *two*. In *Manchester*, under *five*. In *Norwich*, under *five*. In *Northampton*, under *ten* (*a*).—But it appears from *Graunt*'s (*b*) accurate account of the births, weddings, and burials in three country parishes for 90 years; and also, from Dr. *Short*'s collection of observations in his *Comparative History*, and his Treatise entitled, *New Observations on Town and Country Bills of Mortality*; that in country villages and parishes, the major part live to mature age, and even to marry. In the parish of *Holy-Cross* (*c*), in *Salop*, it appears

---

(*a*) See the Tables at the end of this work.

(*b*) See *Natural and Political Observations on the Bills of Mortality*, by Capt. *John Graunt*, F. R. S.—See also Mr. *Derham*'s *Phisico-Theology*, p. 174, where it appears, that in the parish of *Aynho* in *Northamptonshire*, tho' the *births* had been, for 118 years, to the marriages as 6 to 1; yet the *burials* had been to the marriages only as $3\frac{1}{4}$ to 1.

(*c*) This parish contains in it a village which is a part of the suburbs of *Shrewsbury*. It consists of 1400 acres of arable and pasture land; besides 300 acres taken up by houses and gardens. It is six miles in circumference; half of which lies along the banks of the river *Severn*.—I mention these particulars to shew, that it may be reckoned a *country* parish; tho', perhaps, not perfectly so, on account of its nearness to *Shrewsbury*.—The christenings in it exceed the burials a little; and the number of inhabitants

pears from a curious register, which has been kept by the Rev. Mr. *Gorsuch*, the vicar, that, of 655 who have died there at all ages for the last 20 years, 321, or near one half, have lived to 30 years of age: And, by forming a Table of Observations from this register, in the manner which will be described in the last Essay, I find that a child just born in this parish has an expectation of 33 years; and that, in general, under the age of 50, the *expectations* of lives here exceed those in *London*, in the proportion of about 4 to 3.—In the parish of *Ackworth, Yorkshire*, mentioned in the note, p. 184, it appears, from an exact account kept by Dr. *Lee*, of the ages at which all died there for 20 years, or from 1747 to 1767, that *half* the inhabitants live to the age of 46—In the province of *Vaud, Switzerland*, consisting of

habitants (mostly labouring people) has, for the last 20 years, kept nearly to 1050, without any considerable increase.—The register of this parish from 1750 to 1760, has been published in the LIId volume of the *Philosophical Transactions*, Part I. Art. 25. And a continuation of it from 1760 to 1770, has been lately communicated and read to the Royal Society. It is kept with particular care and accuracy by Mr. *Gorsuch*; and furnishes very useful *data* for determining the difference in value between town and country lives.—It deserves to be mentioned particularly, that no *foreigners* or *strangers*, who happen to die in this parish, or who may be brought into it to be buried, are entered into the register: Nor are any of the fixed inhabitants omitted, tho' carried out to be buried.

112,951

112,951 inhabitants, half live to 41.—So great is the difference between the duration of human life in *towns* and in the *country*.—Further evidence for the truth of this obfervation may be deduced from the account given by Dr. *Thomas Heberden*, and publifhed in the Philofophical Tranfactions (Vol. LVII. p. 461), *of the increafe and mortality of the inhabitants of the ifland of Madeira*. In this ifland, it feems, the weddings have been to the births, for 8 years, from 1759 to 1766, as 10 to 46.8; and to the burials, as 10 to 27.5, or 9 to 24.75. Double thefe proportions, therefore, or the proportion of 20 to 46.8, and of 18 to 24.75, are the proportions of the number marrying annually, to the number born and the number dying. Let one marriage in three be a 2d or (*a*) 3d marriage on the fide of either the man or the woman; or, in other words, let one in fix of all that marry be *widows* and *widowers*; and 9 marriages will imply 15 perfons who have grown up to maturity, and lived to marry once or oftener; and the proportion of the number marrying annually the firft time, to the number dying annually, will be 15 to 24.75, or 3 to 5. It may feem to

(*a*) This proportion is taken from fact.—In all *Pomerania*, during 9 years, from 1748 to 1756, the number of perfons who married was 56,956; and of thefe, 10,586 were *widows* and *widowers*. *Sufmilch*'s Works, Vol. I. Tables, p. 98.

follow from hence, that in this island three-fifths of those who die have been married; and, consequently, that only two-fifths of the inhabitants die in childhood and celibacy; and this would be a just conclusion were there no increase, or had the births and burials been equal. But it must be remembered, that the general effect of an increase while it is going on in a country, is to render the proportion of persons marrying annually, to the annual deaths, *greater*, and to the annual births *less*, than the true proportion marrying, out of any given number born. This proportion generally lies between the other two proportions, but always nearest to the first (*a*); and, in the present case, it cannot be so little as one half. Agreeably to this, it appears also from Dr. Heberden's

---

(*a*) In a country where there is no increase or decrease of the inhabitants, and where also life, in its first periods, is so stable, and marriage so much encouraged, that half of all who are born live to be married, the *annual* births and burials must be equal, and also *quadruple* the number of weddings, after allowing for 2d and 3d marriages. Suppose in these circumstances (every thing else remaining the same) the *probabilities of life*, during its first stages, to be improved. In this case, more than *half* the born will live to be married, and an increase will take place. The births will exceed the burials, and both fall below *quadruple* the weddings; or, which is the same, below *double* the number annually married.—Suppose next (the *probabilities of life* and the *encouragement to marriage* remaining the same) the *prolifickness* only of the mar-

Heberden's account, that the expectation of a child just born in *Madeira* is about 39 years; or marriages to be improved. In this case it is plain, that an increase also will take place; but the *annual* births and burials, instead of being less, will now both rise above *quadruple* the weddings; and therefore the proportion of the born to that part of the born who marry (being by supposition two to one) will be less than the proportion of either the *annual* births or the *annual* burials, to the number marrying *annually*.—Suppose again (the *encouragement to marriage* remaining the same) that the *probabilities of life* and the *prolifickness of marriages* are both improved. In this case, a more rapid increase will take place, or a greater excess of the births above the burials; but at the same time they will keep nearer to *quadruple* the weddings, than if the latter cause only had operated, and produced the same increase.—I should be too minute and tedious, were I to explain these observations at large. It follows from them, that, in every country or situation where, for a course of years, the *burials* have been either *equal to* or *less* than the *births*, and both under *quadruple* the marriages; and also that, wherever the burials are *less* than quadruple the annual marriages, and at the same time the births *greater*, there the major part of all that are born live to marry.

I have shewn how the allowance is to be made for 2d and 3d marriages. Very wrong conclusions will be drawn if this allowance is not made. But it is, in part, compensated by the natural children which are included in the births, and which raise the proportion of the births to the weddings higher than it ought to be, and therefore bring it nearer to the true proportion of the number born *annually*, to those who marry annually, after deducting those who marry a 2d or 3d time.

In drawing conclusions from the proportion of *annual births* and *burials*, in different situations, some writers on the increase of mankind, have not given due attention to the difference in these proportions, arising from the different circumstances of increase or decrease among a people.

or more than double the expectation of a child just born in *London*. For the number of inhabitants was found, by a survey made in the beginning of the year 1767, to be 64,614. The annual medium of *burials* had been, for eight years, 1293; of *births* 2201. The number of inhabitants, divided by the annual medium of *burials*, gives 49.89; or the *expectation* nearly of a child just born, supposing the *births* had been 1293, and constantly equal to the *burials*, the number of inhabitants remaining the same. And the same number, divided by the annual medium of *births*, gives 29.35; or the *expectation* of a child just born, supposing the burials 2201, the number of births and of inhabitants remaining the same. And the true *expectation* of life must be somewhere near the mean between 49.89 and 29.35.

people. One instance of this I have now mentioned; and one further instance of it is necessary to be mentioned. The proportion of *annual* births to weddings has been considered as giving the true number of children derived from each marriage, taking all marriages one with another. But this is true only when, for many years, the births and burials have kept nearly equal. Where there is an excess of the births occasioning an increase, the proportion of *annual* births to weddings must be less than the proportion of children derived from each marriage; and the contrary must take place where there is a decrease.

Again:

Again: A 50th part of the inhabitants of *Madeira*, it appears, die annually. In *London*, I have shewn, that above twice this proportion dies annually. In smaller towns a smaller proportion dies (*a*); and the births also come nearer to the burials.—In general; there seems reason to think that in towns (allowing for particular advantages of situation, trade, police, cleanliness, and openness, which some towns may have,) the excess of the burials above the births, and the proportion of inhabitants dying annually, are more or less as the towns are greater or smaller. In *London* itself, about 160 years ago, when it was scarcely a fourth of its present bulk, the births were much nearer

---

(*a*) In *London*, this proportion is, at the highest, 1 in $20\frac{3}{4}$.—In *Norwich*, 1 in $24\frac{1}{4}$.—In *Northampton*, 1 in $26\frac{2}{3}$. See the last Essay. In the parish of *Newbury, Berks*, consisting of 3732 persons, all *town* inhabitants, the annual medium of deaths for 19 years, or from 1747 to 1765, has been 136. In this town, therefore, 1 in $27\frac{1}{2}$ die annually. The contiguous parish of *Speen* consisted, in 1757, of 1200 inhabitants, about 520 of whom were inhabitants of that part of the town of *Newbury* which is in this parish, and the rest were *country* inhabitants. For 34 years, or from 1724 to 1757, *thirty-nine* died here *annually*; or 1 in 31.—In both these parishes the births and burials are nearly equal.—I believe these facts may be depended on; and they seem to shew us very distinctly the gradations in the degrees of human mortality from *great* towns to *moderate* towns, and from *moderate* towns to *small* towns, and to parishes, consisting partly of town and partly of country inhabitants. The next note will shew what the degree of human mortality is in places purely country.

to the burials, than they are now. But in country parishes and villages, the births almost always exceed the burials; and I believe it seldom happens that more than a 40th (*a*) part of the inhabitants die annually. In the four provinces of *New-England* there is a very rapid increase of the inhabitants; but, notwithstanding this, at *Boston*, the capital, the inhabitants would

---

(*a*) According to *Graunt*'s account of a parish in *Hampshire*, not reckoned, he says, remarkably healthful, a 50th part of the inhabitants had died annually for 90 years. *Natural and Political Observations*, &c. Chap. xii.—In the parish of *Ackworth* already mentioned, one of 47 die annually. In the province of *Vaud*, Switzerland, one in 45 die annually. See page 195, and the first part of the *Supplement*. In 1098 country parishes, mentioned by *Sufmilch*, the annual average of deaths, for six years, ending in 1749, was 5255. The number of inhabitants was 225,357. One, therefore, in 43 died annually.—In 106 other parishes, mentioned by him, this proportion was 1 in 50.

In the dukedom of *Wurtemberg*, the inhabitants, Mr. *Sufmilch* says, are numbered every year; and from the average of five years, ending in 1754, it appeared that, taking the towns and country together, 1 in 32 died annually.— In another province, which he mentions, consisting of 635,968 inhabitants, 1 in 33 died annually. From these facts he concludes, that, taking a whole country in *gross*, including all cities and villages, mankind enjoy among them about 32 or 33 years each of existence. And this, very probably, may not be far from the truth in the present state of most of the kingdoms of *Europe*. And it will follow, that a child born in a country parish or village, has, at least, an expectation of 36 or 37 years; supposing the proportion of *country* to *town* inhabitants to be as $3\frac{1}{4}$ to 1; which, I think, this ingenious writer's observations prove to be nearly the case in *Pomerania, Brandenburgh*, and some other kingdoms.

decrease, were there no supply from the country: for, if the account I have seen is just, from 1731 to 1762, the burials all along exceeded the births (*a*). So remarkably do towns, in consequence of their unfavourableness to health, and the luxury which generally prevails in them, check the increase of countries.

Healthfulness and prolifickness are, probably, causes of increase seldom separated. In conformity to this observation, it appears from comparing the births and weddings, in countries and towns where registers of them have been kept, that in the former, marriages, one with another, seldom produce less than four children each; generally between four and five, and sometimes above five. But in towns seldom above four; generally between three and four; and sometimes under three (*b*).

(*a*) See a particular account of the births and burials in this town from 1731 to 1752 in the *Gentleman's Magazine* for 1753, p. 413.

(*b*) Any one may see what evidence there is for this, by consulting Dr. *Short*'s two books already quoted, and the *Abridgment of the Philosophical Transactions*, Vol. VII. part iv. p. 46, and *Graunt*'s account already quoted, of the births, weddings, and burials in three country parishes for 90 years; compared with similar accounts in towns. In considering these accounts, it should not be forgotten that allowances must be made for the different circumstances of increase or decrease in a place, agreeably to the observation at the end of the note in page 196.

I have

I have sometimes heard the great number of old people in *London* mentioned, to prove its favourableness to health and long life. But no observation can be more erroneous. There ought, in reality, to be more old people in *London*, in proportion to the number of inhabitants, than in any smaller towns; because at least one quarter of its inhabitants are persons who come into it from the country, in the most robust part of life, and with a much greater probability of attaining to old age, than if they had come into it in the weakness of infancy. But, notwithstanding this advantage, there are much fewer persons who attain to great ages in *London*, than in most other places where observations have been made.—At *Breslaw* it appears, by Dr. *Halley*'s Table, that 41 of 1238 born, or a 30th part, live to be 80 years of age. The same, I am informed, is true of *Manchester* (a).—In the parish of *All-saints*, in *Northampton*, an account has been kept ever since 1733 of the ages at which the inhabitants die; and I find that a 22d part die there turned of 80. At *Norwich* a like account has been kept; and it appears, that for the last 30 years, a

---

(a) The account I have here given of *Manchester*, and also in page 193, 187, 184, I owe to the information of Dr. *Percival*, a very ingenious and able physician in this town, and author of the *Essays Medical and Experimental*, lately published.

27th part of the inhabitants have died, turned of the fame age.——According to Mr. *Kerffeboom's* Table of Obfervations, publifhed at the end of the third edition of Mr. *De Moivre's* Treatife on the Doctrine of Chances, a 14*th* part die turned of 80. And this is the very proportion that died turned of 80 in the parifh of *Ackworth*, for the 20 years, mentioned page 194. In the parifh of *Holy-Crofs*, already mentioned, p. 184 and p. 193, the *eleventh* part of the inhabitants live to 80 (*a*). See Table III. Supplement. —But in *London*, for 30 years, ending at the year 1768, only 25 of every 1000, who have died, or a 40th part, have lived to this age; which may be eafily difcovered, by dividing the fum of all who have died during thefe years at all ages, by the fum of all who have died above 80.

Among the peculiar evils to which great towns are fubject, I might further mention

---

(*a*) This, however, will appear itfelf inconfiderable, when compared with the following account: "In 1761 the burials in the diftrict of *Chriftianna*, in *Norway*, amounted to 6,929 and the chriftenings to 11,024. Among thofe who died, 394, or 1 in 18, had lived to the age of 90; 63 to the age of 100, and feven to the age of 101.—In the diocefe of *Bergen*, the perfons who died amounted only to 2,580, of whom 18 lived to the age of 100; one woman to the age of 104, and another woman to the age of 108."

See the *Annual Regifter* for 1761, p. 191.

the PLAGUE. Before the year 1666, this dreadful calamity laid *London* almost waste once in every 15 or 20 years; and there is no reason to think, that it was not generally bred within itself. A most happy alteration has taken place; which, perhaps, in part is owing to the greater advantages of cleanliness and openness, which *London* has enjoyed since it was rebuilt; and which lately have been very wisely improved.

The facts I have now taken notice of are so important that, I think they deserve more attention than has been hitherto bestowed upon them. Every one knows that the strength of a state consists in the number of people. The encouragement of population, therefore, ought to be one of the first objects of policy in every state; and some of the worst enemies of population are the luxury, the licentiousness, and debility produced and propagated by great towns.

I have observed that *London* is now (a) increasing. But it appears, that, in truth, this

---

(a) This increase is greater than the bills shew, on account of the omission in them of the two parishes which have been most increased by new buildings; I mean *Marybone* and *Pancrass* parishes. The former of these parishes is now one of the largest in *London*. The annual medium of burials in it for the last 10 years has been 732.—In *Pancrass* parish this medium, for the same time,

this is an event more to be dreaded than desired. The more *London* increases, the more the rest of the kingdom must be deserted; the fewer hands must be left for agriculture; and, consequently, the less must be the plenty, and the higher the price of all the means of subsistence. *Moderate* towns, being seats of refinement, emulation, and arts, may be public advantages. But *great* towns, long before they grow to half the bulk of *London*, become checks on population of too hurtful a nature, nurseries of debauchery and voluptuousness; and, in many respects, greater evils than can be compensated by any advantages (*a*).

Dr.

time, has been 309.—It will, perhaps, be a satisfaction to some to be further informed, that, from an accurate account taken in *March* 1772, it appeared, that the number of inhabitants in that part of this last parish which joins to *London* was then 3479, of whom 1594 were *lodgers*; and that the number of houses was 476, of which about 330 have been built within these seven years.—It will be observed here, that, in this part of *Pancrass* parish, there are above seven persons to a house; but it should be observed likewise, that it consists chiefly of *lodging-houses*, and that the account was taken at a time of the year when it was fullest of *lodgers*; and that, consequently, no conclusion can be drawn from hence with respect to the proportion of inhabitants to houses in *London* in general.

(*a*) The mean annual *births*, *weddings*, and *burials* in the following towns, for some of the last years, have been nearly,

At

Dr. *Heberden* observes that, in *Madeira*, the inhabitants double their own number in 84 years. But this (as you, Sir, well know) is a very slow increase, compared with that which takes place among our colonies in AMERICA. In the back settlements, where the inhabitants apply themselves entirely to agriculture, and luxury is not known, they double their own number in 15 years; and all thro' the northern colonies, in 25 years (*a*). This is an instance of increase so rapid, as to have scarcely any parallel. The births in these countries must exceed the burials much more than in *Madeira*; and a greater proportion of the born must reach maturity.—In 1738, the number of inhabitants in *New Jersey* was taken by order of the go-

	Births.	Weddings.	Burials.
At Paris,	19,100	4,400	19,400
Vienna, from 1757 to 1769	5,800		6,600
Amsterdam,	4,600	2,400	8,000
Copenhagen,	2,700	886	3,300
Berlin, for 5 years, ending at 1759.	3,855	980	5,054

It deserves notice, that before 1770, all that died in the hospitals at *Vienna* were omitted in the bills.—Of the *Paris* bills a more particular account will be given in the Postscript to this Essay.

(*a*) See a Discourse on *Christian Union*, by Dr. *Styles*, *Boston*, 1761, p. 103. 109, &c.—See also, *The Interest of Great Britain considered with regard to her Colonies, together with Observations concerning the Increase of Mankind, peopling of Countries*, &c. p. 35. 2d edit. *London*, 1761.

vernment, and found to be 47,369. Seven years afterwards, the number of inhabitants was again taken; and found to be increased, by procreation only, above 14,000; and very near one *half* of the inhabitants were found to be under (*a*) 16 years of age. In 22 years, therefore, they must have doubled their own number, and the births must have exceeded the burials 2000 annually. As the increase here is much quicker than in *Madeira*, we may be sure that a smaller proportion of the inhabitants must die annually. Let us, however, suppose it the same, or a 50th part. This will make the annual burials to have been, during these seven years, 1000; and the annual births 3000; or an 18th part of the inhabitants.—Similar observations may be made on the much quicker increase in *Rhode Island*, as related in the preface to the *Collection of the London Bills of Mortality*; and also in the valuable pamphlet last quoted, on *the Interest of Great Britain with regard to her Colonies*, p. 36.—What a prodigious difference must there be, between the vigour and the happiness of human life in such situations, and in such a place as *London?*—The original number of persons who, in 1643, had settled in *New-England*, was 21,200. Ever since, it is reckoned, that more have

(*a*) According to Dr. *Halley*'s Table, the number of the living under 16, is but a *third* of all the living at all ages.

left

left them than have gone to them (*a*). In the year 1760, they were increased to half a million. They have, therefore, all along doubled their own number in 25 years. And if they continue to increase at the same rate, they will, 70 years hence, in *New-England* alone, be four millions; and in all *North America*, above twice the number of inhabitants in *Great Britain* (*b*).—But I am wandering from my purpose in this letter. The point

(*a*) See Dr. *Styles*'s pamphlet, just quoted, p. 110, &c.

(*b*) The rate of increase, supposing the procreative powers the same, depends on two causes: The "encouragement to marriage;" and the "*expectation* of a child just born." When one of these is given, the increase will be always in proportion to the other. That is; As much *greater* or *less* as the *ratio* is of the numbers who reach maturity, and of those who marry, to the number born, so much *quicker* or *slower* will be the increase.—Let us suppose the operation of these causes such, as to produce an annual excess of the *births* above the *burials*, equal to a 36th part of the whole number of inhabitants. It may seem to follow from hence, that the inhabitants would double their own number in 36 years; and thus some have calculated. But the truth is, that they would double their own number in much less time. Every addition to the number of inhabitants from the births, produces a proportionably greater number of births, and a greater excess of these above the burials; and if we suppose the excess to increase annually at the same rate with the inhabitants, or so as to preserve the *ratio* of it to the number of inhabitants always the same, and call this *ratio* $\frac{1}{r}$, the period of doubling will be, the *quotient* produced by dividing the logarithm of 2 by the *difference* between the logarithms of $r + 1$ and r; as might be easily demonstrated. In the present case, r being 36, and $r + 1$ being

the State of London, *Population*, &c. 209

point I had chiefly in view was, the prefent ftate of *London* as to healthfulnefs, number of

being 37, the period of doubling comes out 25 years. If *r* is taken equal to 22, the period of doubling will be 15 years.—But it is certain that this ratio may, in many fituations, be greater than $\frac{1}{22}$; and, inftead of remaining the fame, or becoming lefs, it may *increafe*, the confequence of which will be, that the period of doubling will be fhorter than this rule gives it.—According to Dr. *Halley*'s Table, the number of perfons between 20 and 42 years of age is a third part of the whole number living at all ages. The prolific part, therefore, of a country may very well be a 4th of the whole number of inhabitants; and fuppofing four of thefe, or every other marriage between perfons all under 42, to produce *one* birth every year, the annual number of births will be a 16th part of the whole number of people. And, therefore, fuppofing the burials to be a 48th part, the annual excefs of the births above the burials will be a 24th part, and the period of doubling 17 years.—The number of inhabitants in *New-England* was, as I have faid from Dr. *Styles*'s pamphlet, half a million in 1760. If they have gone on increafing at the fame rate ever fince, they muft be now 640,000; and it feems to appear that in fact they are now more than this number. For, fince writing the above obfervations, I have feen a particular account, grounded chiefly on furveys lately taken with a view to taxation, and for other purpofes, of the number of males, between 16 and 60 in the four provinces. According to this account, the number of fuch males is 218,000. The whole number of people, therefore, between 16 and 60, muft be nearly 436,000. In order to be more fure of avoiding excefs, I will call them only 400,000. In Dr. *Halley*'s Table, the proportion of all the living under 16 and above 60, to the reft of the living, is 13.33 to 20; and this will make the number of people now living in the four provinces of *New-England* to be 666,000. But on account of the rapid increafe, this proportion muft be

P   con-of inhabitants, and its influence on population. The obfervations I have made may, perhaps, help to fhew, how the moft is to be made of the lights afforded by the *London* bills; and ferve as a fpecimen of the proper method of calculating from them. It is indeed extremely to be wifhed, that they were lefs imperfect than they are, and extended further. More parifhes round *London* might be taken into them; and, by an eafy improvement in the parifh regifters now kept, they might be extended through all the pa-

---

confiderably greater in *New-England*, than that given by Dr. *Halley*'s Table. In *New Jerfey*, I have faid the number of people under 16, was found to be almoft equal to the number above 16. Suppofe, however, that in *New-England*, where the increafe is flower, the proportion I have mentioned is only 16 to 20; and then the whole number of people will be 720,000.

I cannot conclude this note without adding a remark to remove an objection which may occur to fome in reading Dr. *Heberden*'s account of *Madeira*, to which I have referred. In that account 5945 is given as the number of children under feven in the ifland, at the beginning of the year 1767. The medium of annual births, for eight years, had been 2201; of burials 1293. In fix years, therefore, 13,206 muft have been born; and if, at the end of fix years, no more than 5945 of thefe were alive, 1210 muft have died every year. That is; almoft all the burials in the ifland, for fix years, muft have been burials of children under feven years of age. This is plainly incredible; and, therefore, it feems certain, that the number of children under feven years of age muft, through fome miftake, be given, in that account, 3000 or 4000 too little.

rishes and towns in the kingdom. The advantages arising from hence would be very considerable. It would give the precise law according to which human life wastes in its different stages; and thus supply the necessary *data* for computing accurately the values of all *life-annuities* and *reversions*. It would, likewise, shew the different degrees of healthfulness of different situations, mark the progress of population from year to year, keep always in view the number of people in the kingdom, and, in many other respects, furnish instruction of the greatest importance to the state. Mr. *De Moivre*, at the end of his book on the Doctrine of Chances, has recommended a general regulation of this kind; and observed, particularly, that at least it is to be wished, that an account was taken, at proper intervals, of all the living in the kingdom, with their ages and occupations; which would, in some degree, answer most of the purposes I have mentioned.—But, dear Sir, I am sensible it is high time to finish these remarks. I have been carried in them far beyond the limits I at first intended. I always think with pleasure and gratitude of your friendship. The world owes to you many important discoveries; and your name must live as long as there is any knowledge of philosophy among mankind. That you may ever enjoy all that can make you most happy, is the sincere wish of,

SIR,

Your much obliged,

and very humble Servant,

Newington-Green,
April 3, 1769.

RICHARD PRICE.

POSTSCRIPT.

AT *Edinburgh*, bills of mortality, of the same kind with those in *London*, have been kept for many years. I have, since the foregoing letter was written, examined these bills, and formed a Table of Observations from them, as I found them for a period of 20 years, beginning in 1739, and ending in 1758.—As this is a town of moderate bulk, and seems to have a particular advantage of situation; I expected to find the probabilities of life in it, nearly the same with those at *Breslaw*, *Northampton* and *Norwich*; but I have been surprized to observe, that this is not the case. During the period I have mentioned, only one in 42 of all who died at *Edinburgh*, reached 80 years of age; which is a smaller proportion than attains to the same age in *London*. See p. 203.—In general; it appears, that the probabilities of life in this town are much the same, thro' all the stages of life, with those in *London*, the chief difference being, that after 30, they are rather lower at *Edinburgh*.—It is not difficult to account for this. It affords, I think, a striking proof of the pernicious effects arising from uncleanliness, and crouding together on one spot too many inhabitants. At *Edinburgh*, Mr. *Maitland* says, " the build-
" ings, elsewhere called *houses*, are denomi-
" nated

"nated *lands*; and the *apartments*, in other
"places named *stories*, here called *houses*, are
"so many freeholds inhabited by different
"families; whereby the houses are so ex-
"cessively crouded with people, that the
"inhabitants of this city may be justly pre-
"sumed to be more numerous than those of
"some towns of *triple* its dimensions." See
Maitland's *History of Edinburgh*, p. 140.

In the year 1748, the whole number of *apartments* or *families* in the city and liberties of *Edinburgh*, was 9064. This Mr. *Maitland* mentions as the result of particular examination, and undoubtedly right. *Ib.* p. 217, 218.—In 1743, an accurate account was taken, by the desire of this writer, of the number of *families* and *inhabitants* in the parish of St. *Cuthbert*. *Ib.* p. 171. The number of *families* was 2370, and of *inhabitants* at all ages, 9731. The proportion, therefore, of *inhabitants* to *families*, was $4\frac{1}{10}$ to 1; and, supposing this the true proportion for the whole town, the number of inhabitants will be $4\frac{1}{10}$ multiplied by 9064, or 37,162.— The yearly medium of deaths in the town and liberties for eight years, from 1741 to 1748, was 1783. *Ib.* p. 220 and 222. And, consequently, *one* in $20\frac{4}{7}$ died annually.

Mr. *Maitland*, tho' possessed of the *data* from which these conclusions necessarily followed, has made the number of inhabitants 50,120, in consequence of a disposition to

exag-

exaggerate in these matters, and of assuming, without any reason, a 28th part of the inhabitants as dying annually.

In page 220, he expresses much surprize at finding, that the number of males in this town was less than the number of females, in the proportion of 3 to 4. But this is by no means peculiar to *Edinburgh*.

All I have been saying must be understood of the state of *Edinburgh*, before the year 1758. The bills, for the last 12 years, have been so irregular, and so different from the same bills for the preceding years, and from all other bills, that I cannot give them any credit. Either some particular incorrectness has crept into the method of keeping them; or there has been some change in the state of the town which renders them of no use. Probably the former is the truth.

From the note in p. 206, it appears, that the christenings and burials at PARIS, come very near to equality. This once led me to suspect, that there must be some particular singularity in the state of *Paris*, which rendered it much less prejudicial to health and population than great towns commonly are. But better information has lately obliged me to entertain very different sentiments.—The difference between the births and burials at *Paris*, is much greater than the bills shew. " Children here are baptized the instant " they

"they are boon; and, in a day or two af-
"terwards, it is the cuſtom to ſend them to
"the adjacent villages to be nurſed. A
"great number, therefore, of the infants born
"at *Paris* die in the country, and theſe
"appear only in the regiſter of chriſten-
"ings." See a book entitled the *Police
of France*, page 127. And *Buffon*'s Natural
Hiſtory, Tom. II. at the end.—"All the
"children alſo received into the *Foundling-
"Hoſpital*, are immediately ſent to be nurſ-
"ed in the country, at a diſtance from *Paris*,
"where they remain 5 or 6 years; at the end
"of which time they are brought again to
"*Paris*, the boys to be placed in the ſuburbs
"of St. *Antoine*, and the girls at *Salpetriere*,
"to be further maintained 'till they arrive at
"the age of twelve years." *Police of France*,
p. 81.—The following paſſage in the ſame
writer, containing a further account of this
Hoſpital, is important; and therefore, tho'
long, I cannot help tranſcribing it.—"Let
"us ſuppoſe, that out of 4000 children an-
"nually carried into the country, two thirds
"may die, during the five years they are
"deſtined to remain at nurſe; ſo that only
"1333 would conſtantly be the annual
"number ſent back to *Paris*; who, being
"kept at the two Hoſpitals St. *Antoine* and
"*Salpetriere* juſt mentioned, 'till they are 12,
"and ſucceeded by a like number each year,
"the total number compoſed of all brought
"in

" in the succeffive years, would make the
" constant resting stock to amount to 9331.
" But of these we will suppose a 5th part
" to die every year. Yet even then the
" constant resting stock of children ought to
" be 7465. How greatly then must we be
" surprized to find, by the authentic account
" taken from their own books, only 640
" boys in the college of St. *Antoine*, and not
" more than 600 girls at the *Saipetriere*;
" so that the resting stock of returned found-
" lings appears to be no more than 1240,
" which being deducted from 7465, will make
" the difference in the deficiencies 6225.
" What then becomes of these?—Are they
" reclaimed by their parents?—Or do they
" perish for want of care?—In answer to
" which questions it was explained to me;
" that as many of the lower class of people
" were induced to marry, in order to be ex-
" cused from serving in the militia; so when
" these have children, which they are un-
" able to maintain, they usually send them to
" this hospital; which, therefore, must be
" looked upon, as not only a charity for the
" care of exposed and deserted children whose
" parents are unknown, but also as a public
" *nursery* for the sustenance of the children
" of poor people, who, tho' registered at the
" office, are often reclaimed from their coun-
" try nurses by their parents. This accounts
" in some measure, for the small stock of
" children

"children brought back to the hospital at
"Paris.—The further difference is suspected
"to be owing to the insufficient nourishment
"they receive; as this particular charity, as
"well as the General Hospital, adopts that
"preposterous method of taking in an un-
"limited number, while there is only a li-
"mited income for their subsistence." *Ib.*
page 83.

These facts prove, that, at the same time that the register of *christenings* at *Paris* must be full, the register of *burials* must be very deficient. Let the deficiencies be reckoned at 3700; and, consequently, the annual burials at 23,100. The annual average of weddings, given in p. 206, is 4400; and, therefore, the number of persons who marry annually must be 8800. Deduct a 6th part (*a*) for *widows* and *widowers*, and 7134 will be the number of *virgins* and *batchelors* marrying annually.—The difference between the christenings and burials is 4000; which, therefore, is the number of annual recruits from the country. These, in general, must be persons in mature life. Suppose 3000 of them to marry after settling at *Paris*. Then, 7134 lessened by 3000, or 4134 will be the number of persons born at *Paris* who grow up to marry; and 14,966, or near four-fifths of all who are born at *Paris*, will be the number dying annually in childhood and celibacy.

(*a*) Vid. Note, p. 195.

The

The fuppofitions on which I have made this computation feem moderate; but if any one thinks otherwife, he may make the fame calculation on any other fuppofitions.

The births at *Paris* are above four times the weddings; and it may feem, therefore, that here, as well as in the moft healthy country fituations, every wedding produces above four children. I have obferved nothing like this in any other great town. Many children born in the country are, I fuppofe (*a*), brought to the Foundling-Hofpital, and there chriftened. This Hofpital may likewife occafion a more than common number of illegitimate births. And, befides, fome who leave the country to fettle at *Paris*, may come thither already married. Thefe are circumftances that will fwell the regifter of births, without having any effect on the weddings. I do, however, know that any of them take place at *Paris*; and, perhaps, it muft be granted, that it is diftinguifhed in this refpect from moft other towns. Nor can I wonder at this, if it be indeed true, not only, that all married men in *France* are excufed ferving in the militia from whence draughts are made for the army, but alfo,

---

(*a*) " If the parents of a child brought to this Hofpital are known, the regifter of its baptifm muft be produced. If the parents are unknown, the child muft be baptifed after being received," *Police of France*, page 82.

that

that a *fifth* of all the children born at *Paris* are sent to the *Foundling-Hospital* (*a*). These

(*a*) See the Police of France, p. 83.—This writer adds, that a *third* of all that die at *Paris* die in Hospitals. "In the *Hotel Dieu* (a great Hospital, situated in the "middle of the city) we may, he says, behold a horrid "scene of misery; for, the beds being too few for the "numbers admitted, it is common to see 4, or 6, or "even 8 in a bed together, lying 4 at one end, and 4 "at the other, ill of various distempers in several de- "grees; some bad, others worse; some dying, others "dead.—Above a *fifth* of all admitted to this Hospital "die; the annual numbers admitted being 21,823. The "medium of deaths for three years from 1751 to 1753, "4650.—The medium of deaths for the same years in "*all* the Hospitals was 6181." *Ib.* p. 85.—In our two great city Hospitals, St. *Thomas*'s and St. *Bartholomew*'s, about 600 die annually; or one in 13 of all admitted as in-patients.——An account of the *Hotel Dieu* at *Paris*, much the same with that now given, may be found in the *Memoirs of the Year Two Thousand Five Hundred* lately published, and translated from the French by *W. Hooper*, M.D. "A citizen or stranger (this writer says) who "falls sick, and is sent thither, is imprisoned in a noisome "bed, between a corpse and a person expiring in agonies, "to breathe the noxious vapours from the dead and the "dying, and convert a simple indisposition into a cruel "disease.—Six thousand wretches are crouded together "into this Hospital, where the air has no free circula- "tion; and the arm of the river which flows by, re- "ceives all its filth, and is drank, abounding with the "seeds of corruption, by half the city." The *London* Hospitals, it appears, have greatly the advantage; but indeed, with respect to Hospitals in general, *as now con-structed and regulated*, I cannot help fearing that they cause more distempers than they cure, and destroy more lives than they save. See *Thoughts on Hospitals*, by Mr. *Aikin*, surgeon, together with a Letter to the Author, by Dr. *Percival*.

are

are encouragements to marriage that no other city enjoys. It has been seen that the *Foundling-Hospital*, tho' attended with this effect, is, probably, in the highest degree pernicious; but it is to be wished, that some policy of the same kind with that *first* mentioned, was pursued in this kingdom.—At the end of the 2d vol. of Monsieur *De Buffon's Natural History*, there are Tables formed from the Observations of M. *Du Pre de S. Maur*, of the *French Academy*, containing an account of the ages at which 13,189 persons died in three parishes at *Paris*; and also, of the ages at which 10,805 persons died in 12 country parishes and villages near *Paris*.—According to these Tables, many *more* die in the beginning of life, and much *fewer* in the latter part of life, in the country than in *Paris*. But the circumstances of *Paris*, and the country round it, are such, that no argument can be drawn from hence in favour of *Paris*. Many of the children dying in the country, are children sent thither from *Paris* to be nursed; and, on the other hand, *many*, perhaps *most*, of those who die in old age at *Paris*, are persons who have removed thither from the country, some to *Hospitals*, and some to places and settlements. It is evident, therefore, that these Tables give a representation of the probabilities of life at *Paris*, which, when compared with those in the adjacent country,

country (*a*), is juft the reverfe of the truth. Were the children born at *Paris*, who die in the country, to be transferred to the town regifter; and, on the contrary, the adults born in the country, who die at *Paris*, to be transferred to the country regifter, there is no reafon to doubt, but that the probabilities of life at *Paris*, would be found as low, in comparifon with thofe in the country, as the probabilities of life in *London* are; or, perhaps, much lower.—This obfervation is applicable, in fome degree, to moft other great towns; and, in general, on account of the migrations from the country to towns, navies and armies, we may be fatisfied, that we err on the fide of *defect*, whenever we judge of the probabilities of life in the *country*, from the numbers dying in the feveral ftages of life; and, on the fide of *excefs*, whenever, in the fame way, we judge of the probabilities of life in *towns*. And this, it is obvious, has a tendency to confirm all that has been faid in the preceding Effay, concerning the pernicious effects of great towns on human life.

There are feveral *ordonnances* and *arrets* of council which fix the boundaries of *Paris*,

---

(*a*) It is for this reafon that thefe Tables, when combined, exhibit juftly the *mean* probabilities of life for town and country taken together; and that the Table of the *decrements* of life deduced from them by M. *Buffon* and Mr. *Du Pre*, agrees nearly with Dr. *Halley*'s Table.

and prohibit all new buildings beyond those boundaries.—The reasons of this regulation, as set forth in one of these *arrets*, are remarkable; and it will not be improper to recite them.—" By the excessive aggrandiz-
" ing of the city, it is said, the air would be
" rendered unwholesome, and the cleaning
" the streets more difficult."—" Augment-
" ing the number of inhabitants would aug-
" ment the price of provisions, labour, and
" manufactures."—" That ground would be
" covered with buildings which ought to be
" cultivated in raising the necessary subsist-
" ence for the inhabitants; and thereby ha-
" zard a scarcity."—" The people in the
" neighbouring towns and villages would be
" tempted to come and fix their residence in
" the capital, and desert the country."—
" And lastly; the difficulty of governing so
" great a number of people, would occasion
" a disorder in the *Police*, and give an oppor-
" tunity to rogues to commit robberies and
" murders (*a*)."

No one can think overgrown cities greater evils than I do. But, yet, I can by no means approve of this policy. The effect of it must be, crouding together too many people within the prescribed boundaries, and rendering a town more the seat of uncleanliness, infection and disease.—The number of houses in

(*a*) Vid. Police of France, p. 130.

Paris

Paris is reckoned about 28,000 (*a*), but the number of inhabitants, (suppoſing a 20th part to die annually, and the true number of burials to be 23,000) muſt be 460,000; or about 16 times the number of houſes.

It is happy for LONDON, that there have been no laws to reſtrain its increaſe. In conſequence of being allowed to extend itſelf on all ſides into the country, the inhabitants now take near twice the room to live upon that they did; and it is become leſs the means of ſhortening human life. See p. 191, 192, and 204.

In page 206, I have given the annual *medium* of births, weddings and burials at BERLIN, from 1755 to 1759.—In 1747, an account was taken with the utmoſt care, by the order of the King of PRUSSIA, of the

---

(*a*) Vid. Police of France, p. 130.
I find, in a Book entitled, *Recherches ſur la Population des Generalites d'Auvergne, de Lyon, de Rouen*, &c. by M. MESSANCE, and printed at *Paris* in 1766, the number of houſes at *Paris* is given 23,565, from a capitation tax in 1755; and the number of families 71,114. There muſt, I ſuppoſe, be ſome deficiencies in this account; but M. *Meſſance*, by allowing moſt extravagantly (See Note, p. 183.) 8 to a family, infers from it that the number of inhabitants at *Paris* is 568,912.—On very unſatisfactory grounds alſo he makes the inhabitants of FRANCE to be near 24 millions. *Suſmilch* calls them 16 millions; and *Marſhal Saxe*, in his Memoirs on the Art of War, after obſerving that *Vauban*'s calculation had made them 20 millions; adds, that their number at the time he wrote was far inferior to this.

number of inhabitants in this town; and, it was found to be 107,224.—In order to be more certain, a *second* account was taken the same year; and the number found the same within 200.—In 1755, the inhabitants were increased to 126,661. Their number, therefore, in 1758, could scarcely be less than 134,000; and must have been to the annual burials nearly as $26\frac{1}{4}$ to 1.—This proportion is higher than could be expected in a town so considerable; and also so much crouded, as to have, at an average, 16 inhabitants in every house. But an observation already made, must be here remembered. —BERLIN, for many years, had been increasing very fast, by a conflux of people from the surrounding country and provinces. About the year 1700, the medium of annual burials was no more than 1000. In 50 years, therefore, it has more than quadrupled itself.—In a city increasing with such rapidity, the *ratio* of inhabitants to the annual deaths, must be greatly above the just standard.— Were there now, such accessions to LONDON of deserters from the country, in the beginning of mature life, as would cause the number of inhabitants to increase at the rate of 10,000 every year, it would in 60 years be doubled; and the proportion of inhabitants to deaths would rise gradually, 'till it came to be about one-third greater. BERLIN, we have seen, has, in fact, increased at *more than*

than *double* this rate; and, therefore, the number of inhabitants dying annually in it is in reality very high.

The ingenious *Sufmilch*, to whose works, I owe my information concerning BERLIN, makes the proportion of people who die annually in *great* towns, to be from $\frac{1}{24}$ to $\frac{1}{27}$; in *moderate* towns, from $\frac{1}{28}$ to $\frac{1}{32}$; and in the country from $\frac{1}{40}$ to $\frac{1}{50}$.—The observations and facts in this Essay, joined to those which will be found in the 4th Essay and the *Supplement*, prove, I think, that these proportions may be more truly stated as follows.—*Great* towns, from $\frac{1}{19}$ or $\frac{1}{20}$ to $\frac{1}{23}$ or $\frac{1}{27}$. *Moderate* towns, from $\frac{1}{23}$ to $\frac{1}{27}$. The *country*, from $\frac{1}{35}$ or $\frac{1}{40}$, to $\frac{1}{50}$ or $\frac{1}{60}$.—This, however, must be understood with exceptions. There may be *moderate* towns so ill situated, or whose inhabitants may be so crouded together, as to render the proportion of deaths in them greater than in the largest towns: And, of this, EDINBURGH, if it is not now, was 20 years ago an example.—There may be also *great* towns in which, from a sudden increase, this proportion may be less than in small towns: And of this I have just given an example in BERLIN.

ESSAY

ESSAY II.

On Mr. DE MOIVRE's *Rules for calculating the Values of* Joint *Lives.*

THE calculation of the values of *single* and *joint* lives, from given Tables of Observation, being tedious and troublesome; Mr. *De Moivre* has had recourse to two *Hypotheses*, which give easy rules for this purpose; and which, he thought, corresponded with sufficient exactness to Observations.— The first of these *Hypotheses* is, that the probabilities of life decrease, as we advance from childhood to old age, in an *arithmetical progression*; or in such a manner, that the *difference* is always the same, between the number of persons living at the beginning of any one year, and the number living at the beginning of the next following year.—The other *Hypothesis* is, that the probabilities of life decrease in a *geometrical* progression; or in such a manner, that the *proportion* is always the same, between the number of persons living at the beginning of any one year, and the number living at the beginning of the next following year.—All the Tables of Observation shew, that the real law, according to which human life wastes, comes much nearer to the former *Hypothesis*, than the latter.—In Tables III, IV, and V, in the *Appendix*, it is so near the former *Hypothesis*, that the difference is scarcely worth regarding. According to this *Hypothesis*, therefore, (accommodated to the *Breslaw* Table, in the manner mentioned in the note, page 2.) Mr. *De Moivre* calculated the values of *single* lives; and the rules founded upon it for this purpose are so easy, that an operation which would otherwise take up much time, may be performed almost immediately.

By proceeding on the same principles, the values of *joint* lives might have been calculated; but the rules for this purpose derived from these principles, are far from being equally easy in practice. Here, therefore, Mr. *De Moivre* quitted his *first* Hypothesis; and finding, that the *second* Hypothesis afforded, in the case of *joint* lives, rules that were as easy, as the rules given by the other Hypothesis were in the case of *single* lives, he chose to adopt this *Hypothesis*; believing at the same time, that the values of *joint* lives, obtained by rules derived from it, would not deviate much from the truth. But in this he was greatly mistaken. The values of *two joint* lives obtained by these rules are so wrong, that in finding the present value, in a *single payment*, of one life after another, they generally give results which are near a *quarter* of the true value too great; and about *two-fifths*

fifths too great, when the value is sought in *annual payments* during the joint lives.— These are errors so considerable, that I think it of particular importance that the public should be informed of them, in order to prevent the inconveniencies and perplexities they may occasion.

Mr. *Simpson* (in the Appendix to his Treatise on the *Doctrine of Annuities and Reversions*) has observed, that Mr. *De Moivre*'s rules for finding the values of joint lives are wrong. But I don't know, that it has been ever attended to, that they are *so* wrong as I have found them. Mr. *Simpson*'s remarks point out chiefly the errors in these rules, when the values of *three* or more joint lives are calculated by them; but, 'till I was forced to a particular examination of this subject by some difficulties into which I found myself brought by following Mr. *De Moivre* too implicitly, I did not at all suspect, that any such errors as I have mentioned, could arise from these rules, when the values of only *two* joint lives are calculated by them. Mr. *De Moivre*, in consequence of other remarks contained in Mr. *Simpson*'s *Appendix*, altered, in the 4th edition of his Treatise, some of his rules. It is surprizing he did not see reason at the same time to alter these.

That there may be no doubt about the truth of these observations, I will just mention a few examples of the difference between
the

the values of a given reverfionary annuity, according to the rules to which I have objected, and the *true* values, according to the exact method of deducing them from Mr. *De Moivre's firft hypothefis.*

Let the propofed annuity be 30 *l.*; to be enjoyed for what fhall happen to remain of the life of a perfon now 40 years of age, after the life of another perfon of the fame age. The value of the joint lives (intereft being at 4 *per cent.*) is, by problem 2d of Mr. *De Moivre's* Treatife on Life-Annuities, 8.964; which fubtracted from 13.196, (the value by Table VI, of a fingle life at 40) gives 4.23; which remainder, multiplied by 30, gives *l.* 126.9, or the value of the reverfion in a fingle prefent payment. And 126.9, divided by the foregoing value of the joint lives, is *l.* 14.16; or, the value of the reverfion in annual payments during the joint lives.—But the *true* values are *l.* 101.1 in a *fingle* payment, by Queft. I. chap. I.; and *l.* 10.3, in *annual* payments, by Queft. IV.— The former values, therefore, are a *quarter* of the true value too great in the *fingle* payment; and near *two-fifths* too great in the *annual* payments.

The *true* value of the fame annuity for a life at 66, after another life of the fame age, is, (reckoning intereft as before, at 4 *per cent.*) 68 *l.* in a *fingle* payment; and 13.5 in *annual* payments.—But thefe values, according to the Problem juſt quoted, are 91 *l.* and 21 *l.* one of which is near a *third,* and the other above *half* the true value too great.

In *unequal* lives theſe errors may be no leſs conſiderable.—Thus; if the value of the propoſed annuity be required for a life at 70, after a life at 30 years of age; it will, by the ſame Problem, be *l.* 26.5, in a *ſingle* payment; and *l.* 5.1, in *annual* payments during the joint lives. But the *true* values are 17 *l.* and *l.* 3.05.

Where 3 or more lives are concerned the errors will be ſtill greater.

The true values of the joint lives, mentioned in theſe Examples, have been calculated by a rule in page 16, of Mr. *Simpſon*'s Treatiſe on the Doctrine of *Annuities* and *Reverſions*, and explained in note (M) *Appendix.*—To ſave, however, a great deal of trouble hereafter, I have thought proper to calculate Table VII, which gives the exact values according to Mr. *De Moivre's firſt* hypotheſis, of two joint lives, for every five years of human life, from 10 to 70.

This *hypotheſis,* I have obſerved, does not differ much from the Tables of Obſervation in the *Appendix,* for *Breſlaw, Northampton* and *Norwich.* Between the ages of 30 and 40, it gives the values of *ſingle* lives almoſt the ſame with the *Breſlaw* Table. Under 30, it gives them ſomewhat *leſs;* and above 40, ſomewhat *greater.* But it ought to be remembered, that wherever it does this, it gives, at the same ages, the values of the *joint* lives also too little or too great; and that, consequently, the results from it, in calculating the values of *Reversions*, and of the *longest* of given lives, come so much nearer to exactness.

The rules to which I have objected are the only ones given by Mr. *De Moivre*, in all the editions of his Treatise on Life-Annuities. But it seems, this great mathematician became at last sensible, that they were too incorrect; and, therefore, at the end of the last edition of his Treatise on the *Doctrine of Chances*, page 320, (a work which gets into comparatively few hands) he has given other rules which come nearer the truth. But even these rules produce errors so great in many cases, (particularly when combined with the errors of the hypothesis) that it will be best never to use them.

ESSAY

ESSAY III.

Of the Method of calculating the Values of Reverſions depending on Survivorſhips.

ALL Queſtions relating to the values of lives and reverſions, are at preſent of particular importance in this kingdom. Much buſineſs is continually tranſacted in this way; and any conſiderable errors in the methods of ſolving ſuch queſtions, muſt in time produce very bad conſequences.—The deſign of the following obſervations is to point out a particular error, into which there is danger of falling, in finding the values of ſuch reverſions as depend on ſurvivorſhips. In doing this, I ſhall, in order to be as plain as poſſible, take the following caſe. "A, aged "40, expects to come to the poſſeſſion of "an eſtate, ſhould he ſurvive B, aged like-"wiſe 40. In theſe circumſtances he offers, "in order to raiſe a preſent ſum, to give ſe-"curity for 40 *l. per annum*, out of the eſtate "at his death, provided he ſhould get into "poſſeſ-

"possession; that is, provided he should sur-
"vive B. What is the sum that ought now
"to be advanced to him, in consideration of
"such security, reckoning compound inte-
"rest at 4 *per cent.* ?"

Mr. *De Moivre's* directions in his Treatise on Annuities, Problems 17th and 20th, lead us to seek the required sum in this case, by the following process.

Find first, the present sum A should receive, for the reversion of 40 *l. per annum* for ever after his death; supposing it *not* dependent on his surviving B. The present value of such a reversion is "the (*a*) value of the life "subtracted from the *perpetuity*, and the re-"mainder multiplied by the annual rent."— The value of the life is, by Table VI, 13.196. This subtracted from 25, the *perpetuity*, leaves 11.80; which, multiplied by 40, gives *l.* 472; the value of the supposed estate, after the life of A. But, as Mr. *De Moivre* observes, the lender having a chance to lose his money, a compensation ought to be made to him for the risk he runs, which is founded on the possibility, that a man of 40 years of age may not survive another person of the same age. This chance is an *equal* chance; and, therefore, half the preceding sum, or 236 *l.* is the money which should be advanced now on the expectation mentioned.

(*a*) By *Scholium*, p. 34, and Problem 26th, p. 293, of Mr. *Simpson's* Select Exercises.

This folution carries a plaufible appearance; and moft perfons will, probably, be ready to pronounce it right; nor will this be at all wonderful, as fo great a mafter of thefe fubjects as Mr. *De Moivre* appears to have been mifled by it.—Nothing more is neceffary to prove it to be fallacious, than proceeding in the fame way to folve the following fimilar Queftion.

" A, aged 40, offers to give fecurity for
" 40 *l. per annum*, to be entered upon at his
" death, provided it fhould happen *before* the
" death of B, aged likewife 40. What fum
" fhould now be advanced to him for fuch
" a reverfion, intereft being reckoned at 4
" *per cent.*?"

In folving this Problem, agreeably to the method juft defcribed, we are to find the value of 40 *l. per annum*, to be entered upon *certainly* at the death of A; and then to multiply this value by the chance that A fhall *not* furvive B, or by $\frac{1}{2}$; and in this way the anfwer comes out the fame with that already given.

Now it may be eafily feen, that this muft be wrong. The value of a reverfion, to be received when a perfon of a given age dies, cannot be the fame, whether the condition of obtaining it is, that he fhall die *before*, or that he fhall die *after* another perfon. That is, whether it is provided, that a purchafer, if he fucceeds, fhall get into poffeffion *fooner* or *later*,

later. The reverſion in the latter caſe muſt, without doubt, be of leſs value than in the former.

The firſt Queſtion here propoſed, reſolves itſelf into the following general Queſtion.

"What is the preſent value of a given re-verſionary eſtate, to be entered upon after the failure of two lives, provided one *in particular* of them ſhould be the *longeſt life?*"

Now, the preſent value of an eſtate to be enjoyed for ever, after the failure of the *longeſt* of two lives, is "the value of the *longeſt* of the two lives, ſubtracted from the *perpetuity*; and the remainder multiplied by the annual rent of the eſtate."—The value of the *longeſt* of two lives is (as is well known) the value of the two *joint* lives, ſubtracted from the *ſum* of the (*a*) values of the two *ſingle* lives. In the preſent caſe, therefore, it is 9.82, (the value of two joint lives at the age of 40 by Table VII,) ſubtracted from twice 13.196; (the value of a *ſingle* life at the ſame age by Table VI,) that is, 16.57 year's purchaſe. And this ſubtracted from 25, (the perpetuity) gives 8.43; which, multiplied by 40, gives *l*. 337.2, the value of the given eſtate were it *certainly* to be enjoyed, after the ex-

---

(*a*) See Mr. *De Moivre* on Annuities, Problem IV; or Mr. *Simpſon's Doctrine of Annuities and Reverſions*, Problem II.

tinction of the longest of two lives both 40; that is, whether *one* or *other* of them failed last. But that A's life in particular should fail last, rather than B's, is an even chance. The true value of the reversion, therefore, is half the last value, or *l.* 168.6.

In like manner. The second Question is the same with the Question, "What is the pre-
"sent value of 40 *l. per ann.* for ever, to be en-
"tered upon after the extinction of two *joint*
"lives both 40; that is, whenever *either* of
"them shall fail; provided the first that fails
"should happen to be A's life in particular?"
—And the answer is found by subtracting the present value of the *two joint* lives from the *perpetuity*, and multiplying the remainder by $\frac{1}{2}$, or by the chance that A in particular shall die first: And this will give the required value, *l.* 303.4 (*a*).

In short. It appears in *both* these cases, that, according to the first method of solution, we are to subtract from the *perpetuity* the value of *one* of the single lives; when, in the *former* case, the value of the *longest* of the two lives, and, in the *latter* case, the value of their *joint continuance*, ought, in reality, to be subtracted. I need not say what prodigious errors may often arise from hence; and how unfit such a method of solution is for practice.

(*a*) I have, tho' scarcely necessary, given a demonstration of these Solutions in the Appendix, note (N).

Mr.

Mr. *Simpson*, in p. 322, of his Select Exercises, speaks on this subject in the following manner.—" I have been very particular on these kinds of Problems; and the more so, as there has been no method before published, that I know of, by which they can be rightly determined. 'Tis true, the manner of proceeding, by first finding the probability of survivorship, (which method is used in my former work, and which a celebrated author has largely insisted on in three successive editions) may be applied to good advantage, when the given ages are nearly equal; but then it is certain, that this is not a genuine way of going to work, and that the conclusions hence derived are at best but near approximations."

This excellent mathematician has here expressed himself much too favourably of the method of solution on which I have remarked.—In both the cases I have specified, the ages are equal; and yet, in one of them the error is a good deal above a *third* of the true value, and in the other a *fifth:* And, it is obvious, that in cases where three equal lives are taken, the errors will be much greater.—Mr. *Simpson*'s Observations in this passage are true only, when applied to a *different* method used by himself, in the 28th and following Problems of his Treatise on the *Doctrine* of *Annuities* and *Reversions*. This method is exact when the lives are equal; but,

it

it gives refults which are too far from the truth, when there is any confiderable inequality between the lives.

It is with reluctance I have made fome of thefe remarks. Mr. *De Moivre* has made very important improvements in this branch of fcience; and the higheft refpect is due to his name and authority. This, however, only renders thefe remarks more neceffary.

In the firft Chapter (Queftions 10th, 11th, 12th, 14th, &c.) I have given a minute account of the method of finding, in all cafes, the values of the reverfions which have been the fubject of this Effay.

ESSAY

ESSAY IV.

Observations on the proper Method of constructing Tables for determining the Rate of human Mortality, the Number of Inhabitants, and the Values of Lives in any Town or District, from Bills of Mortality in which are given, the Numbers dying annually at all Ages.

IN every place that just supports itself in the number of its inhabitants, without any recruits from other places; or where, for a course of years, there has been no increase or decrease, the number of persons dying every year at any particular age, and above it, must be equal to the number of the living at that age.—The number, for example, dying every year, at all ages, from the *beginning* to the utmost *extremity* of life, must, in such a situation, be just equal to the whole number *born* every year. And for the same reason, the number dying every year at *one* year of age and upwards; at *two* years of age and upwards; at *three* and upwards, and so on; must be equal to the numbers that attain to those ages every year; or, which is the

the same, to the numbers of the living at those ages. It is obvious, that unless this happens, the number of inhabitants cannot remain the same. If the former number is greater than the latter, the inhabitants must *decrease;* if less, they must *increase.*—From this observation it follows, that in a town or country where there is no increase or decrease, bills of mortality which give the ages at which all die, will shew the exact number of inhabitants; and also the exact law, according to which human life wastes in that town or country.

In order to find the number of inhabitants; the mean numbers dying annually, at every particular age and upwards, must be taken as given by the bills, and placed under one another in the order of the second column of the 12th Table in the Appendix. These numbers will, it has appeared, be the numbers of the living at 1, 2, 3, &c. years of age; and, consequently, the sum, diminished by half the number born annually (*a*), will be the whole

---

(*a*) This subtraction is necessary for the following reason.—In a Table formed in the manner here directed, it is supposed, that the numbers in the second column are all living together at the beginning of every year. Thus; the number in the *second* column opposite to 0 in the *first* column, the Table supposes to be all just born together on the first day of the year. The number, likewise, opposite to 1, it supposes to attain to one year of

R age

whole number of inhabitants.—In such a series of numbers, the excess of each number above that which immediately follows it, will be the number dying every year, out of the particular number alive at the beginning of the year; and these excesses set down regularly as in the third column of the Table to which I have referred, will shew the different rates at which human life wastes thro' all its different periods, and the different probabilities of life at all particular ages.

It must be remembered, that what has been now said goes on the supposition, that the place, whose bills of mortality are given, supports itself, by procreation only, in the number of its inhabitants. In towns this very seldom happens, on account of the luxury and debauchery which generally prevail in them. They are, therefore, commonly kept up by a constant accession of strangers or *settlers*,

age just at the same time that the former number is born. And the like is true of every number in the second column.—During the course of the year, as many will die at all ages as were born at the beginning of the year; and, consequently, there will be an excess of the number alive at the beginning of the year, above the number alive at the end of the year, equal to the whole number of the annual births; and the true number constantly alive together, is the arithmetical mean between these two numbers; or, agreeably to the rule I have given, the *sum* of the numbers in the second column of the Table, lessened by *half* the number of annual births. See Essay I, page 174.

who

who remove to them from country parishes and villages. In these circumstances, in order to find the true number of inhabitants, and probabilities of life, from bills of mortality containing an account of the ages at which all die; it is necessary that the proportion of the annual births to the annual settlers should be known; and also the period of life at which the latter remove.—Both these particulars may be discovered in the following method.

If for a course of years there has been no sensible increase or decrease in a place, the number of annual settlers will be equal to the excess of the annual burials above the annual births. If there is an *increase*, it will be *greater* than this excess. If there is a *decrease*, it will be *less*.

The period of life at which these settlers remove, will appear in the bills by an increase in the number of deaths at that period and beyond it. Thus; in the *London* bills, the number of deaths, between 20 and 30, is generally above double, and between 30 and 40, near triple the number of deaths between 10 and 20: And the true account of this is, that from the age of 18 or 20, to 35 or 40, there is an afflux of people every year to *London* from the country, which occasions a great increase in the number of inhabitants at these ages; and, consequently, raises the deaths for all ages *above* 20, considerably

siderably above their due proportion, when compared with the number of deaths *before* 20.—This is obfervable in all the bills of mortality for towns with which I am acquainted, not excepting even the *Breflaw* bills. Dr. *Halley* takes notice, that thefe bills give the number of deaths, between 10 and 20, too fmall. This he confidered as an irregularity in them, owing to chance; and, therefore, in forming his Table of Obfervations, he took the liberty fo far to correct it, as to render the proportion of thofe who die to the living in this divifion of life, nearly the fame with the proportion which, he fays, he had been *informed* (*a*) die annually of the young lads in *Chrift-Church Hofpital*. But the truth is, that this irregularity in the bills was derived from the caufe I have juft affigned.—During the five years for which the *Breflaw* bills are given by Dr. *Halley*, the births did, indeed, a little exceed the burials; but, it appears, that this was the effect of fome peculiar caufes that happened to operate juft at that time; for, during a complete century from 1633 to 1734, the annual medium of *births* was 1089 (*b*), and of bu-

---

(*a*) See *Lowthorp*'s Abridgment of the Philofophical Tranfactions, vol. III. p. 670.—Dr. *Halley*'s information in this inftance was not right, as will appear prefently; and, therefore, he has by no means fufficiently corrected the irregularity I have mentioned.

(*b*) See Dr. *Short*'s Comparative Hiftory, p. 63.

rials 1256 (*a*). This town, therefore, must have been all along kept up by a number of yearly recruits from other places, equal to about *a seventh* part of the yearly births.

What has been now observed concerning the period of life at which people remove from the country to settle in towns, would appear sufficiently probable, were there no such evidence for it as I have mentioned; for it might be well reckoned, that these people in general, must be single persons in the beginning of mature life, who, not having yet obtained settlements in the places where they were born, migrate to towns in quest of employments.

Having premised these Observations, I shall next endeavour to explain distinctly, the effect which these accessions to towns must have, on Tables of Observation formed from their bills of mortality. This is a subject proper to be insisted on, because mistakes have been committed about it; and because also, the discussion of it is necessary to shew, how near to truth the values of lives come as deduced from such Tables.

(*a*) It appears from the account in the *Philosophical Transactions*, (Abridgment, vol. VII, No. 380, p. 46, &c.) that from 1717 to 1725, the annual medium of births at *Breslaw* was 1252, of burials 1507; and also, that much the greatest part of the births died under 10 years of age.—From a Table in *Susmilch*'s works, Vol. I. p. 38, it appears, that, in reality, the greater part of all that die in this town are children under five years of age.

The following general rule may be given on this subject.

If a place has, for a course of years, been maintained in a state nearly stationary, as to number of inhabitants, by recruits coming in every year, to prevent the decrease that would arise from the excess of the burials above the births; a Table formed on the principle, "that the number dying annually, after every particular age, is equal to the number living at that age," will give the number of inhabitants and the probabilities of life, too *great* for all ages preceding that at which the recruits cease; and after this, it will give them *right*.—If the accessions are so great as to cause an *increase* in the place, such a Table will give the number of inhabitants and the probabilities of life, too *little*, after the age at which the accessions cease (*a*); and too great, if there is a decrease. *Before* that age it will in *both* cases give them too great; but most considerably so in the former case, or when there is an increase.

(*a*) Agreeably to these Observations; if a place increases, not in consequence of accessions from other places, but of a constant excess of the births above the deaths; a Table, constructed on the principle I have mentioned, will give the probabilites of life too low through the *whole extent* of life; because, in such circumstances, the number of *deaths* in the *first* stages of life must be too great, in comparison of the number of deaths in the *latter* stages; and more or less so, as the *increase* is more or less rapid. The contrary, in all respects, takes place where there is a decrease, arising from the excess of the *deaths* above the *births*.

For

For example. Let us suppose, that 244 of those born in a town, attain annually to 20 years of age; and that 250 more, all likewise 20 years of age, come into it annually from other places; in consequence of which, it has, for a course of years, been just maintained in the number of its inhabitants, without any sensible increase or decrease. In these circumstances, the number of the living in the town of the age of 20, will be always 244 *natives* and 250 *settlers*, or 494 in all; and, since these are supposed all to die in the town, and no more recruits are supposed to come in; 494 will be likewise the number dying annually at 20 and upwards.—In the same manner; it will appear on these suppositions, that the number of the living, at every age, subsequent to 20, will be equal to the number dying annually at that age and above it; and, consequently, that the number of inhabitants and the decrements of life, for every such age, will be given exactly by the Table I have supposed. But for all ages *before* 20, they will be given much too great. For let 280 of all born in the town, reach 10. In this case, 280 will be the true number of the living in the town, at the age of 10; and the recruits not coming in 'till 20, the number given by the bills, as dying between 10 and 20, will be the true number dying annually of the living in this division of life. Let this number be 36; and it will follow,

follow, that the Table ought to make the numbers of the living at the ages between 10 and 20, a feries of decreafing means between 280 and (280 diminifhed by 36, or) 244. But in forming the Table on the principle I have mentioned, 250 (the number above 20 dying annually in the town who were not born in it) will be added to each number in this feries; and, therefore, the Table will give the numbers of the living, and the probabilities of life in this divifion of life, almoft twice as great as they really are.—This obfervation, it is manifeft, may be applied to *all* the ages under 20.

It is neceffary to add, that fuch a Table will give the number of inhabitants, and the probabilities of life, equally wrong before 20, whether the recruits all come in at 20, agreeably to the fuppofition juft made, or only *begin* then to come in. In this laft cafe, the Table will give the number of inhabitants, and probabilities of life, too great throughout the whole extent of life, if the recruits come in at all ages above 20. But if they ceafe at any particular age, it will give them right only from that age; and before, it will err all along on the fide of excefs; but lefs confiderably between 20 and that age, than before 20.——For example. If, of the 250 I have fuppofed to come in at 20, only 150 then come in, and the reft at 30; the numbers of the living will be given 100 too high,

at

at every age between 20 and 30; but, as juſt ſhewn, they will be given 250 too high at every age before 20.—In general, therefore, the number of the living *at* any particular age, muſt be given by the ſuppoſed Table, as many too great as there are annual ſettlers *after* that age: And, if theſe ſettlers come in at all ages indiſcriminately, during any certain interval of life; the number of inhabitants and the probabilities of life will be continually growing leſs and leſs wrong, the nearer any age is to the end of that interval. —Theſe Obſervations prove, that Tables of Obſervation formed in the common way, from bills of mortality for places, where there is an exceſs of the burials above the births, muſt be erroneous, for a great part of the duration of life, in proportion to the degree of that exceſs. They ſhew likewiſe, at what parts of life the errors in ſuch Tables are moſt conſiderable, and how they may be in a great meaſure corrected.

All this I ſhall beg leave to exemplify and illuſtrate a little further, in the particular caſe of *London*.

The number of deaths, between the ages of 10 and 20, is always ſo ſmall in the *London* bills, that it ſeems certain few recruits come to *London* under 20; or at leaſt not ſo many as before this age are ſent out for education to ſchools and univerſities. After 20, great numbers come in 'till 30, and some perhaps 'till 40 or 50.—But, at every age after 50, it is probable, that more retire from *London* than come to it.—The *London* Tables of Observation, therefore, being formed on the principle I have mentioned, cannot give the probabilities of life right 'till 40. Between 30 and 40 they must be a little too high; but more so between 20 and 30; and most of all so before 20.—It follows also, that these Tables must give the number of inhabitants in *London* much too great.

Table XII, in the Appendix, is a Table formed in the manner I have explained, from the *London* bills for 10 years, from 1759 to 1768; and adapted to a 1000 born as a *radix*. The sum of the numbers in the second column, diminished by half the number born, is 25,757. According to this Table then, for every 1000 deaths in *London*, there are $25\frac{3}{4}$ as many inhabitants; or, in other words, the expectation of a child just born is $25\frac{1}{4}$; and the inhabitants are to the annual burials, as $25\frac{3}{4}$ to 1. —But it has appeared, that the numbers in the second column being given on the supposition, that all who die in *London* were born there, must be too great; and we have from hence a DEMONSTRATION, that the probabilities of life are given in the common Tables of *London* Observations, too high, for, at least, the first 30 years of life; and also, that the number of inhabitants in *London* must be *less* than

than 25¼, multiplied by the annual *burials*.—The common Tables, therefore, of *London* Obfervations, undoubtedly want to be corrected (*a*); and the way of doing this, and in general, the right method of forming genuine Tables of Obfervation for towns, may be learnt from the following rule.

"From the fum of all that die annually,
"*after* any given age, fubtract the number
"of annual fettlers *after* that age; and the
"remainder will be the number of the liv-
"ing *at* the given age."

This rule can want no explication or proof, after what has been already faid.

If, therefore, the number of annual fettlers in a town at every age could be afcertained; a perfect Table of Obfervations might be formed for that town, from bills of mortality, containing an account of the ages at which all die in it. But no more can be learnt in this inftance from any bills, than the whole number of annual *fettlers*, and the general divifion of life in which they enter. This, however, may be fufficient to enable us to form Tables that fhall be tolerably exact.—

(*a*) The ingenious and accurate Mr. *Simpfon* faw that it was neceffary to correct the *London* Tables, and he has done it with great judgment; but, I think, too imperfectly, and without going upon any fixt principles, or fhewing particularly, how Tables of Obfervation ought to be formed, and how far in different circumftances, and at different ages, they are to be depended on.

For

For instance. Suppose the *annual deaths* in a town which has not increased or decreased, to have been for many years, in the proportion of 4 to 3 to the *annual births*. It will hence follow, that ¼ of the persons who die in such a town are *settlers*, or *emigrants* from other places; and not *natives:* And the sudden increase in the deaths after 20, will also shew, agreeably to what was before observed, that they enter after this age. In forming therefore a Table for such a town, a *quarter* of all that die at all ages throughout the whole extent of life, must be deducted from the sum of all that die after every given age before 20; and the remainder will be the true number living at that given age. And if, at 20, and every age above it, this deduction is omitted, or the number of the living at every such age is taken the same with the sum of all that die after it, the result will be (supposing *most* of the settlers to come in before 30, and *all* before 40) a Table exact 'till 20; too high between 20 and 30; but nearly right for some years before 40; and after 40 exact again.—Such a Table, it is evident, will be the same with the Table last described at all ages above 20; and different from it only under 20.—It is evident also that, on account of its giving the probabilities of life too great for some years, after 20, the number of inhabitants deduced from it may be depended on as somewhat greater

greater than the truth; and more or less so, as the annual recruits enter in general later or sooner after 20.

Let us now consider, what the result of these remarks will be, when applied particularly to the *London* bills.

It must be here first observed, that, at least one *quarter* of all that die in *London* are *settlers* from the country, and not *natives*.—The medium of annual burials for 10 years, from 1759 to 1768, was 22,956; of births 15,710. The excess is 7246; or near a *third* of the burials.—The same excess, during 10 years, before 1750, was 10,500; or, near *half* the burials. *London* was then *decreasing*. For the last 12 or 15 years it has been *increasing*. This excess, therefore, agreeably to the foregoing observations, was then *greater* than the number of annual settlers; and it is now *less*. I have chosen, however, to suppose the number of annual settlers to be now no more than a *quarter* of the annual burials, in order to allow for more omissions in the births than the burials; and also, in order to be more sure of obtaining results that shall not exceed the truth.

Of every thousand then who die in *London*, only 750 are *natives*, and 250 are *settlers*, who come to it after 18 or 20 years of age: And, consequently, in order to obtain from the bills a more correct Table than the 12th in the *Appendix*, 250 must be subtracted

tracted from every one of the numbers in the second column, 'till 20; and the numbers in the third column muſt be kept the ſame, the bills always giving theſe right.—After 20, the Table is to be continued unaltered; and the reſult will be, a Table which will give the numbers of the living at all ages in *London* much nearer the truth, but ſtill ſomewhat too high.—Such is the 13th Table in the Appendix.—The ſum of all the numbers in the ſecond column of this Table, diminiſhed by 500, is 20,750. For every 1000 deaths, therefore, in *London*, there are, according to this Table, 20,750 living perſons in it; or for every ſingle death, 20¾ inhabitants. It was before ſhewn, that the number of inhabitants in *London* could not be ſo great as 25 times ¼ the deaths. It now appears, (ſince the numbers in the ſecond column of this Table are too high) that the number of inhabitants in *London* cannot be ſo great as even 20 times ¾ the deaths. And this is a concluſion which, I believe, every one who will beſtow due attention on what has been ſaid, will find himſelf forced to receive. It will not be amiſs, however, to confirm it by the following fact, the knowledge of which I owe to the particular enquiry and kind information of Mr. *Harris*, the ingenious maſter of the Royal Mathematical School in Chriſt-Church Hoſpital:— The average of lads in this ſchool has, for 30 years

years paſt, been 831. They are admitted at all ages between ſeven and eleven; and few ſtay beyond 16. They are, therefore, in general, lads between the ages of eight and 16. They have better accommodations than it can be ſuppoſed children commonly have; and about 300 of them have the particular advantage of being educated in the country. In ſuch circumſtances it may be well reckoned that the proportion of children dying annually, muſt be leſs than the general proportion of children dying annually at the ſame ages in *London*.—The fact is, that, for the laſt 30 years, $11\frac{1}{4}$ have died annually; or *one* in $70\frac{2}{3}$.

According to Table XIII, *one* in 73 dies between 10 and 20, and *one* in 70 between eight and 16. That Table, therefore, probably, gives the decrements of life in *London*, at theſe ages, too little, and the numbers of the living too great: And, if this is true of theſe ages, it muſt be true of all other ages *under* 20; and it follows demonſtrably, in conformity to what was before ſhewn, that more people ſettle in *London* after 20, than the quarter I have ſuppoſed; and that from 20 to at leaſt 30 or 35, the numbers of the living are given too great, in proportion to the decrements of life.

In this Table the numbers in the ſecond column are doubled at 20, agreeably to what really happens in *London*; and the ſum of

the

the numbers in this column diminifhed by half the whole number of deaths, gives the *expectation* of life, not of a child juft born, as in other Tables, but of all the inhabitants of *London* at the time they enter it, whether that be at birth, or at 20 years of age. The *expectations*, therefore, and the *values* of *London* lives under 20, cannot be calculated from this Table. But it may be very eafily fitted for this purpofe by finding the number of births which, according to the given decrements of life, will leave 494 alive at 20; and then adapting the intermediate numbers in fuch a manner to this radix, as to preferve all along the number of the living, in the fame proportion to the numbers of the dead. This is done in the 14th Table in the Appendix; and this Table may, I fancy, be recommended as better adapted to the prefent ftate of *London* than any other Table. The values of lives, however, deduced from it, are in general nearly the fame with thofe deduced by Mr. *Simpfon*, from the *London* bills as they ftood 40 years ago. The main difference is, that after 52, and in old age, this Table gives them fomewhat lower than Mr. *Simpfon*'s Table.

It has fufficiently appeared, what judgment we are to form of the values of lives thus deduced. During the greateft part of the interval of life, in which the annual recruits that keep up *London* come to it, thefe values err on the side of *excess:* and after that interval, they err, perhaps, a little on the side of *defect* (a) on account of retirements from *London* in the last stages of life.

The

(*a*) I have not taken into account the effect of migrations *from* towns, on Tables formed in the manner I have explained; because, towns in general being kept up by recruits from the country, the migrations *from* them are of little consequence, compared with the migrations *to* them.—Thus; in LONDON, it appears from the much greater number of deaths between 40 and 50, than in any other equal interval of life after 10, that more people come to it than leave it, at every age between 20 and 50. After 50, it is probable, that the contrary happens. But, it should be considered, that *emigrants* from LONDON after 50, are chiefly persons who, having got fortunes in business, chuse to leave off, and to spend the latter part of their lives in country retirements. But how few are these compared with the multitudes who, tho' possessed of good fortunes, never retire; and with the bulk of the inhabitants in lower stations, who never can be able, without the greatest inconveniencies, to quit the settlements by which they are supported? It is, however, likely, that retirements from LONDON are now more numerous than they ever were; and that they have *some* effect on the bills of mortality, and on *Tables* formed from them; by causing these Tables to give the number of the living too little, in comparison with the *decrements* of life, at every age, from that at which the migrations *to* and *from* LONDON become equal, to the age at which the latter cease.—To explain this; let us suppose, that none settle in LONDON after 50; but that, between 35 and 50, as many come to it as retire from it at all ages after 35; and that these retirements cease at 70. In this case, the Tables will give the proportion of the living to the decrements of life too high 'till 35. *At* 35, this proportion will be given right. *After* 35, it will begin to

The number of inhabitants in *London* may also be learnt from what has been offered, more nearly than by any method which has been hitherto taken. It cannot, it has been shewn, exceed 20 times ¼ the number of annual deaths. Could, therefore, the annual deaths be ascertained, we should know the number of inhabitants within pretty narrow limits. But the omissions in the bills are such, that it is not possible to ascertain, with exactness, the annual deaths. Dr. *Brakenridge* supposed these omissions to amount to 2000 annually. The result of a very minute enquiry by Mr. *Maitland* is, that in the year 1729, they amounted to 3038. But they are probably now much more considerable, than either of these writers have reckoned

---

to be given too low; and this error will increase 'till 50; from which age it will decrease gradually 'till it vanishes at 70: And after 70, the Tables will be exactly right again.—This is the exact state of the effect of retirements from *London*, on the *London* Table of Observations. But this effect appears, indeed, to be inconsiderable; for, after 50, the values of lives by the *London* Table, are continually approaching nearer and nearer to the same values by other Tables; which could not happen were retirements attended with any *great* effect.—It is proper to add, that in summing up, as above explained, the numbers of the living, in order to find the number of inhabitants in *London*, the circumstance that these numbers may be too small for some years after 40 or 50, in consequence of retirements, is, undoubtedly, much more than balanced by their being given too high between 20 and 40.

them (*a*). Let them be 6000; and the number of inhabitants will be 601,750 at most.

All the preceding Observations are, it is plain, applicable to bills of mortality for towns in general; and point out the way of deducing from them genuine Tables of Observations, which shall give the true probabilities and values of lives, and the true number of inhabitants, in the town whose bills are given.—I shall beg leave to confirm and illustrate this, in the particular case of the town of NORTHAMPTON.

In this town, containing four parishes, namely, *All-Saints, St. Sepulchre's, St. Giles,* and St. *Peter's,* an account has been kept ever since the year 1741, of the number of males and females that have been christened and buried (Dissenters included) in the whole town. And in the parish of *All-saints,* containing the greatest part of the town, an account has been kept ever since 1735, of the *ages* at which all have died there.

In 1746, an account was taken of the number of *houses,* and of *inhabitants* in the town. The number of *houses* was found to be 1083; and the number of *inhabitants* 5136.—In the parishes of *All-Saints* and *St. Giles,* the number of *male* and *female heads* of families, *ser-*

(*a*) Vid. Preface to a Collection of the Bills of Mortality from 1657 to 1758, p. 4, &c.

vants, *lodgers*, and *children*, were particularly distinguished.—The *heads of families* were, 707 males; and 846 females.——*Children*, males 624; females 759.—*Servants*, males 203; females 280.—*Lodgers*, males 137; females 287.—In *St. Peter*'s, males 99; females 129.—In *St. Sepulchre's*, *adults* 638; *children* 427. In this parish the sexes were not distinguished.

The *Christenings* and *Burials* in the *whole town* for 28 years, from 1741 to 1770, have been as follows.

Christened { Males 2361 / Fem. 2288 } 4649—Annual medium 155

Buried { Males 2869 / Fem. 2878 } 5747—Annual medium 191

In the parish of *All-Saints*, from 1735 to 1770, or 36 years,

Christened { Males 1632 / Fem. 1610 } 3242—Annual medium 90

Buried { Males 1856 / Fem. 1834 } 3690—Annual medium 102½

Of these died,

Under 2 years of age	1206
Between 2 and 5	276
Between 5 and 10	155
Between 10 and 20	155
Between 20 and 30	297
Between 30 and 40	257

Between 40 and 50 — 297
Between 50 and 60 — 300
Between 60 and 70 — 293
Between 70 and 80 — 285
Between 80 and 90 — 155
Between 90 and 100 — 14

Total 3690

A Table formed from these *data* in the manner of Table XII; or, on the supposition, that all who *die* in *Northampton* were *born* there, would give the expectation of a child just born 28.83 years; or, the proportion of the inhabitants to the annual deaths, as 28.83 to 1. It has been shewn, that this proportion, in a place where the burials exceed the births, must be *greater* than the *true* proportion of the number of inhabitants to the annual deaths: And this appears to be the real case. For the bills shew, that, from 1741 to 1750, or for 10 years, about the time when the number of inhabitants was 5136, the annual medium of burials was 197.5; which, multiplied by 28.83, gives 5693; or a 9th part more than the *true* number.

A Table formed in the manner of Table XIII, would give the proportion of inhabitants to the annual deaths, as 26.41 to 1; and this makes the inhabitants 5216; or very near the true number.

The IVth Table, in the *Appendix*, is formed in the same manner with Table XIV, for *London:* And this is the genuine Table of Observations for *Northampton*, from which may be calculated the true probabilities and values of lives, at all ages, in that town.

At Norwich, bills of mortality, of the same kind with those in *London* and *Northampton*, have been kept for many years. I have been favoured with a copy of these bills for 30 years, from 1740 to 1769. The annual medium of *christenings*, during this period, has been 1057 (*a*), of *burials* 1206. And from hence, together with the account of the numbers dying in the several decads of life, after 10, I have formed Table V, which shews the true probabilities of life in this town.

(*a*) In this register all that die before baptism, and also all that are born and die among *Quakers*, *Jews*, &c. are omitted. There are also some other omissions; and the true annual medium of births and burials must be greater than they are given in the bills. But this will have no effect on a Table of Observations, supposing the *proportions* of the births to the burials, and of the numbers dying in the different stages of life, given right. —It is proper I should mention further here, that these bills give only the whole number of children dying under 10, without specifying the numbers dying under two years of age, between 2 and 5, and between 5 and 10, as in other bills. I have, therefore, in forming the Table for Norwich, supposed the proportions of these numbers the same that they are at Northampton.

The

The following particulars seem to deserve notice here.

First. Had these Tables been formed from the NORTHAMPTON and NORWICH bills, for no longer time than any 10 years taken together, of the periods I have mentioned; they would have given the values of lives nearly the same. These Tables, therefore, are founded on a sufficient number of Observations; and it appears, that there is an invariable law which governs the waste of human life in these towns.—The same remark might be made concerning LONDON (*a*). See p. 256.

Secondly. An account was taken at SHREWSBURY, in 1750, of the *whole* num-

---

(*a*) Some have entertained a very wrong notion of the imperfections in the LONDON bills. They do, indeed, give the *whole* number of births and deaths much too little; but the conclusions with respect to the probabilities of life in LONDON, and the proportion of inhabitants dying annually, depend only (agreeably to the observation in the last note) on the *proportions* of the numbers dying in the several divisions of life; and these are given *right* in the LONDON bills.—For first, There seems nothing in this case, that can be likely to cause the deficiencies in the bills to fall in one division of life more than in another: But what decides this point is, that these proportions, as given by the bills for *any* ten, or even *any* five years, come out nearly the same with one another; and always very different from the proportions given by other bills.—There are no other variations, than such as must arise from the fluctuations of LONDON, as to increase and decrease; and also from some improvements in its state, which have lately taken place. See Essay I, p. 191, 192, 204.

ber of inhabitants; distinguishing, particularly, the number at the age of 21 and upwards.—The former number was 8141; and the latter, 5187.—According to a Table formed for NORTHAMPTON, in the same manner with Table XIII. for LONDON, the whole number of the living is to the number of the living at 21 and upwards, as 26,411 to 16,586; that is, as 8141 to 5113.—According to a like Table for NORWICH, these numbers are to one another, as 24,500 to 15,680; that is, as 8141 to 5210.—These Tables, therefore, give the proportion of the *whole* number of inhabitants, to the number of the living at 21 and upwards, almost exactly the same with the true proportion, as it is at SHREWSBURY (a): And this affords a kind of demonstration of the rectitude of the principles on which these Tables have been formed.

In the parish of HOLY-CROSS near SHREWSBURY, an account was taken in 1760 and

(a) The annual medium of births at SHREWSBURY, for 7 years, from 1762 to 1768, was 301; of burials 329. It appears, therefore, that one in 24½ of the inhabitants die annually. But it should be remembered, that in 1766, the small-pox and measles increased very much the mortality in this town; and I find also, that, since 1750, a nursery for *foundlings* from LONDON, was established here; and that in 1768 this nursery contained 660 children and servants. It seems, therefore, probable, that the true *medium* of burials about the year 1750, must have been less than 329; and that the proportion of inhabitants dying annually, may not be much greater than it is at NORTHAMPTON; or 1 in 26.41.

1770,

1770, of the *whole* number of inhabitants; diftinguifhing, *both* times, the number at the age of 70 and upwards; and the *laft* time, the number at 10 and upwards: And I find, that a Table formed from the *Regifter* of this parifh, mentioned p. 193, 194, gives, likewife, *thefe* numbers as nearly the fame as could poffibly be expected.

But further.—The number of inhabitants, not reckoning children, in the parifhes of *St. Giles* and *All-Saints*, NORTHAMPTON, was, in 1746, 2460; and the *whole* number of inhabitants in thefe two *parifhes* was 3843. See p. 259.—In the account I have received, the particular age at which the limit of childhood was fixed in taking this furvey, is not mentioned; but there is fufficient reafon to believe, that it was 21: And, taking this for granted, the number of inhabitants, not children, will come out, (by fuch a Table for NORTHAMPTON, as Table XIII for LONDON) 2414; or, nearly the fame with the number *really* found in thefe parifhes.—Had this number been computed, from a Table formed for NORTHAMPTON, in the manner of Table XII, *Appendix*, it would have come out only 2176. This remark is applicable to the Table for *Breflaw*, formed by Dr. *Halley*, compared with the fame Table, corrected for all the ages under 20 (*a*), by the rule, p. 251.

The

(*a*) I have given Dr. *Halley's* Table in the Appendix juft as he framed it. A correction of it might be made

from

The neceffity, therefore, of that correction is verified by facts; and it appears, abundantly, that the Tables I have given for NORTHAMPTON and NORWICH may be depended on.

But, thirdly. In comparing thefe two Tables, it may be obferved, that there is a difference between them in favour of NORTHAMPTON, *fewer* dying there in childhood, and *more* in old age. The fame would be found to be true, were the NORTHAMPTON Table to be compared with a corrected BRESLAW Table. It appears, therefore, agreeably to what might have been expected, that NORTHAMPTON, being a fmall town compared with BRESLAW and NORWICH, is lefs unfavourable to health and longevity. The difference, however, is not confiderable. After the age of 20, there is a ftriking conformity between all the three Tables, which gives them great weight and authority.

Further. It ought to be particularly noted, that thefe Tables prove the *decrements* from the proportion of births to burials, mentioned p. 244. And it would then appear, that a 25th part of the inhabitants at *Breflaw* die annually; and that half the number born die there under fix, as well as at *Norwich*. This Table, as we now have it, makes half live to 16; but the account mentioned in the note, page 245, fhews this not to be the truth. It likewife makes the number of inhabitants at SHREWSBURY, above the age of 21, to be 4730; and in the parifhes of *All-Saints* and *St. Giles*, in NORTHAMPTON, 2230. It gives, therefore, thefe numbers wrong; whereas, as obferved above, a corrected Table would give them true.

of life in moderate towns, to be nearly *equal* thro' most of its stages. At NORTHAMPTON it appears that, of a given number of persons alive at 20, the same number die every year 'till 78, without any interruption worth notice, except between the ages of 30 and 40.—A like uniform decrease in the probabilities of life appears in the BRESLAW and NORWICH Tables; but not so remarkably. It was this circumstance in the BRESLAW Table, that led Mr. *De Moivre* to the *Hypothesis*, described in p. 2, and so often mentioned in this work.—The values of lives, I have said, deduced from this *Hypothesis*, agree so nearly with the same values deduced immediately from the Tables, that it is scarcely worth while to distinguish them. But that every one may be able to judge of this for himself, I have calculated (*a*) the following Table.

Value of a life at the age	By *Breslaw* Table.	By *Norwich* Table.	By *Northampton* Table.	By Mr. *De Moivre's Hypothesis*.
12	17.617	17.48	17.20	16.69
20	16.49	16.41	15.93	15.89
30	14.77	15.15	14.85	14.68
40	12.90	13.36	13.10	13.19
50	10.87	11.13	11.25	11.34
60	8.58	8.54	9.02	9.01
70	5.59	5.99	6.26	6.06
75	4.21	4.86	4.79	4.29

Reckoning interest at 4 per cent.

(*a*) Every calculation of this kind may be made without much labour, by a rule explained in note (O) *Appendix*.

It may be observed in this Table, that the values, by the *Hypothesis*, come nearer to the true values by the NORTHAMPTON and NORWICH Tables, than by the BRESLAW Table; and also, that, before the age of 60, they are *all* much higher than the values for the same ages in LONDON by Table X; the inhabitants of *London*, (as Mr. *De Moivre* observes) being " for causes (*a*) too well known, " more short-lived than the rest of mankind."
—The *Hypothesis*, therefore, is by no means applicable to LONDON lives. It is proper to add, that neither can it be applied to the valuation of COUNTRY lives.—It appears, from the register of the parish of *Holy-Cross* (*b*), that the *expectations* of lives there are much greater than the *expectations* by the *Hypothesis*.
—The *expectation* there of a life (*c*)

At		By *Hypoth.*	In *Lond.*
20 is	38	33	28.9
27	33.9	29.5	25.1
30	32	28	23.6
40	25.7	23	19.6
50	20	18	16
60	14.5	13	12.4
70	10	8	8.8

(*a*) Doctrine of Chances, p. 347.
(*b*) See Essay I. p. 193, 194.—I have in the *Supplement* given the Table of Observations from whence these conclusions are deduced. In p. 263. a fact is mentioned, which seems to prove, that 20 years is a period long enough to afford *data* in this case of sufficient authority. It is, however, certain, that the same register continued 10 or 20 years longer, will afford *data* more to be depended on.

From this comparison it appears, that the *Hypothesis*, from 20 to 60, gives nearly the medium.

(c) The expectation of a child just born in this parish, is 33. At NORTHAMPTON, 25½. At NORWICH, 23¼. In LONDON, 18.—In this parish, 1 in 11 dies at 80, and upwards. In NORTHAMPTON; 1 in 22. In NORWICH; 1 in 27. In LONDON; 1 in 40. See Essay I. p. 202.

I will add, that the probabilities of life here, appear to be much the same, with the probabilities of life among the ministers and professors in SCOTLAND.—This is a fact of some consequence; and, therefore, I shall beg leave to give a brief account of it.

The mean age at which the ministers and professors enter into benefices and professorships in *Scotland*, is reckoned to be 27. Their number is 974. The establishment among them for providing for their widows, begun on the 25th of *March* 1744; from which time to *November* 22, 1770, 774 have died: That is; 29.07 annually; or 1 in 33¼. The *expectation*, therefore, of a life among them, at the age of 27, is 33½; which is nearly the same with the *expectation*, as given above, of a life of the same age in the parish of *Holy-Cross*; and 3¼ years more, than the *expectation* of the same age by Tables III, IV and V. —Now, the *expectation* at a given age, being composed of all the probabilities of life from that age to the extremity of life; there arises from hence reason for concluding, that the probabilities of life among the ministers in *Scotland*, cannot differ much in any part of life, from those in this parish.—But there is another fact that confirms this observation.

The annual average of weddings among the ministers and professors in SCOTLAND, for the last 27 years, has been 31. The average of married persons among them, for 17 years, ending in 1767, had been 667. This number, divided by 31, gives 21½, the *expectation* of marriage among them; which is above 2¼ years more than the *expectation* of marriage would be, by Dr. *Halley*'s Table, on the supposition, that all 1st, 2d and 3d marriages may be justly considered as commencing, one with another, so early as the age of 30.—The *expectation* of two equal

joint

medium between the *expectations* of LONDON and COUNTRY lives; and for this reason it is excellently adapted to general use.—After 60, the *expectations* and *values* of lives in LONDON approach nearer and nearer to the *expectations* and *values* of lives in *Northampton*, *Norwich*, and *Breslaw*; 'till, at 70, they come to be almost the same. This is a circumstance which, I believe, has not been attended to: And it is the more surprising, as there is no cause known, which can produce any error in the values of lives after 60, deduced from the LONDON Table, except migrations from *London*; and the effect of these must be to *diminish* these values.

The following observations will, perhaps, account for this.

It has been proved, that at least *half* the inhabitants of LONDON, turned of 20 years of age, are *emigrants* to LONDON from the

---

joint lives is to the *expectation* of a *single* life of the same age, as 2 to 3, by note (L) *Appendix*. It follows, therefore, that among the ministers in *Scotland*, the *expectation* of a *single* life at 30, cannot be less than 32.25. Most probably it is more; on account of the later commencement of marriage in the situation of the *Scotch* ministers.—I reckon also, that 27 must be less than the mean age at which they enter their benefices and professorships; meaning by it, not the age on each side of which equal numbers enter; but the age at which, the excess of the interval of time taken to enter on one side, is just such as to compensate the greater numbers who enter on the other side. See the conclusion of note (F) *Appendix*.

country.

country. So great a change as that, from the *country* air and modes of life, to the air and modes of life in *London*, muſt be particularly hurtful to theſe perſons; and, therefore, (except *infants*) it is in them, probably, that the pernicious influence of *London* on its inhabitants chiefly takes place. They come in at every age 'till near 50; and this is the reaſon why the deaths continually increaſe in *London* 'till that age; but, after that age, the inhabitants confiſting chiefly of perſons, who (like men *uſed* to drink) have been *ſeaſoned* to *London*, or with whom it does not happen particularly to diſagree; the number of deaths becomes leſs, and the values of lives begin to approach nearer to the common ſtandard in ſmaller towns.

There is one more fact which I ſhall here take notice of; and which deſerves more attention than has been hitherto beſtowed upon it. I mean; "the difference between the "probabilities of life among *males* and *fe-* "*males*, in favour of the latter."

From the account in p. 260, it appears, that at NORTHAMPTON, tho' more *males* are born than *females*, and nearly the ſame number die; yet the number of living *females* is greater than the number of *males*, in the proportion of 2301 to 1770, or 39 to 30. This cannot be accounted for, without ſuppoſing, that *males* are more ſhort-lived than *females*.—One
obvious

obvious reason of this fact is, that *males* are more subject to untimely deaths by accidents of various kinds; and also, in general, more addicted to the excesses and irregularities which shorten life. But this is by no means the *only* reason. For it should be observed, that at NORTHAMPTON the number of *female* children was, in 1746, greater than the number of male children, in the proportion of 759 to 624.—The greater mortality of males, therefore, takes place among *children*.—But this, together with the greater mortality in general of males at all ages, will more particularly appear from the following recital of facts.

In the parish of *Holy-Cross, Salop*, the ingenious Vicar, Mr. *Gorsuch*, in 1760, and again in 1770, took the number of male and female inhabitants turned of 70. In 1760, the number of females turned of this age, was 35; of males, 8. In 1770, these numbers were, females, 35; males, 26. And for the last 10 years, 11 out of 365 have died between the ages of 85 and 102; and they were all females.

At BERLIN, it appeared, from the accurate account which was taken of the inhabitants in 1747, and which has been mentioned in p. 224, 225, that the number of *female* citizens exceeded the number of *male* citizens, in the proportion of 459 to 391: And yet, out of this smaller number of males, more had died for 20 years preceding 1751, in the proportion of 19 to 17 (*a*).

At EDINBURGH, in 1743, the number of *females* was to the number of *males*, as 4 to 3; (See Essay I. p. 215) but the females that died annually, from 1749 to 1758, were to the males, in no higher proportion than $3\tfrac{1}{5}$ to 3. Before 1749, the bills gave the totals of burials, without distinguishing them into the totals of males and females dying every year.

Mr. *Kerseboom*, in his Essay on the numbers of people in HOLLAND, informs us, that from the Tables of assignable Annuities for lives in HOLLAND, which had been kept there for 125 years, wherein the ages of the persons dying are truly entered; it appears, that females have, in all accidents of age, lived about 3 or 4 years longer than the same number of males. See *Philosophical Transactions* abridged, Vol. IX. p. 326.

In Volume the 7th of the *Philosophical Transactions* abridged, Part IV. p. 46, &c. there is an account of the numbers of *male* and *female* still-born children and chrysoms, and of boys and girls under 10, of married men and married women, and of widows and widowers, who died for a course of years at *Vienna, Breslaw, Dresden, Leipsic, Ratisbon,* and some other towns in GERMANY.

(*a*) Vid. *Susmilch, Gottliche Ordnung,* &c. where a minute account is given of the number of males and females at BERLIN in 1747; and also, of the numbers of each sex that had died from 1722 to 1750.

He that will take the pains to examine thefe accounts will find that, though in thefe towns the proportion of males and females born is no higher than 19 to 18, yet the proportion of boys and girls (*a*) that die is 8 to 7; and that, in particular, the *ſtill-born* and *chryſom males*, are to the ſtill-born and *chryſom females*, as 3 to 2.

In thefe accounts it appears alfo, that of 7270 *married* perſons who had died in theſe towns (*b*), 4336 were married *men*, and but 2934 married *women*; that is, *three* married *men* died to two married *women*.—In all POMERANIA, during 9 years, from 1748 to 1756, there died 13,556 married men, and 10,007 married women; that is, nearly 15 to 11. *Suſmilch, Gottliche Ordnung*, vol. i. tables, p.97. The fcheme for making proviſion for the widows and orphans of the miniſters in SCOTLAND, has obliged them to keep an account of the number of *weddings* among them, and the number of *widows* left annually; and it appears, from the *reports* of the *truſtees* for carrying this fcheme into exe-

---

(*a*) In the accounts from *Breſlaw* it is particularly mentioned, that by *boys* and *girls* are meant children to 10 years of age, of whom, for 8 years from 1717 to 1725, *ſeven* males died to *ſix* females, excluſively of the *ſtill-born* and *chryſoms*.

(*b*) In *Breſlaw* alone, for the eight years mentioned in the laſt note, 1891 married men died, to 1196 married women; that is 5 to 3.

cution, that the annual medium of *weddings* (*a*), is (as observed in the note, page 269) 31. And the annual medium of widows, who have come upon the scheme for 27 years, is $19\frac{1}{5}$. Of 31 marriages then contracted annually, $19\frac{1}{5}$ become extinct by the deaths of *husbands*; and but 11.8 by the deaths of *wives*. That is; among the ministers and professors in SCOTLAND, 20 married men die to 12 married women; or 5 to 3. It appears, therefore, that there is the chance of 3 to 2, and in some circumstances even a *greater* chance, that the *woman* shall be the survivor of a marriage, and not the *man*. In order to account for this by the difference of age between men and their wives, this difference ought to be at least 12 years (*b*). That is; supposing the mean age at which women marry to be 23, the mean age at which men marry ought to be 35. But this seems to exceed the bounds of credibility; and, there-

---

(*a*) The annual medium of weddings, among the ministers admitted to benefices, has been, for 27 years from the commencement of the scheme, 27. Besides these, I find there have been 4 weddings annually among them, *before* admission to benefices. The whole annual medium, therefore, is no more than 31.

(*b*) The chance of survivorship between two persons aged 21 and 34, is nearly 3 to 2 in favour of the former. There is the same chance of survivorship between 25 and 37; and 28 and 39. This may be learnt from Problem XVI, in Mr. *De Moivre*'s Treatise on *Life-Annuities*.

fore, very probably, the greater mortality of males muſt operate in this caſe.

It is further obſervable in the accounts from *Germany*, to which I have referred, that the number of *widows* dying annually, is four times the number of *widowers* (*a*); and, as *widows* are certainly, one with another, ſeveral years younger than *widowers*; it may be concluded from hence, that the number of the former in life together could not be leſs than five times the latter.—This fact is likewiſe confirmed, by the obſervations which have been made among the miniſters in *Scotland*. In 1770, the number of *widows* in life, derived from the whole body of miniſters and profeſſors, was 380; but the number of *widowers* among them has, one year with another, been ſcarcely 90; that is, not ſo much as a *quarter* of the number of *widows*.— It may be eaſily ſeen, and it would not be difficult to demonſtrate, that neither the greater number of perſons left widows, nor any pro-

---

(*a*) In *Dreſden* alone, the number of *widows* who died, in four years, was 584. The number of *widowers*, 149. That is; 4 to 1.—At WITTENBERG, during 11 years, 98 *widowers* died, and 376 *widows*.—At GOTHA, during 20 years, 210 *widowers* and 760 *widows*. *Suſmilch's Gottliche Ordnung*, Vol. II. p. 273.—In the country, on account of a leſs difference between the ages of huſbands and wives and more early marriages, the deaths of *widowers* and *widows* come nearer to one another; for in POMERANIA, during the 9 years mentioned in p. 274, the *widowers* that died were 411, the *widows* 1553; or 2 to 5.

bable suppofition concerning the greater frequency of marriages among widowers, can completely account for this, without admitting the *greater mortality of males*.—This, therefore, appears on the whole to be a fact well eftablifhed: And it follows from it, that in order to calculate the values of *Life-Annuities* and *Reverfions* with exactnefs, there ought to be diftinct Tables of the Probabilities of life for *males* and *females*. All that is neceffary to obtain the proper *data* for forming fuch Tables is, that the *fexes* as well as the *ages* of the dead fhould be fpecified in the bills; and this is an improvement of our bills (*a*) of mortality which would give little trouble, and which, therefore, I hope, will be fome time or other made.

It has been obferved, that the author of nature has provided, that more *males* fhould be born than *females*, on account of the particular wafte of *males*, occafioned by wars and other caufes. Perhaps it might have been obferved with more reafon, that this provifion had in view, that particular weaknefs or delicacy in the conftitution of males,

---

(*a*) This improvement would be rendered more complete, by diftinguifhing the *males* that die, under the denominations of *married men*, *widowers*, and *batchelors*; and the *females*, under the denominations of *married women*, *widows*, and *virgins*.—The ufe I have made of fome accounts of this kind which have been kept in *Germany*, fhews that this would be of confiderable fervice.

which makes them more subject to mortality; and which, consequently, renders it necessary, that more of them should be produced, in order to preserve in the world a due proportion between the two sexes.

In the course of this Essay, it has often appeared, that I have been particularly indebted to an information which I have received from NORTHAMPTON.—I should be inexcusable, did I not mention, that I owe this information to Mr. *Lawton,* an ingenious gentleman in that town, who has preserved the bills of mortality there with much care, and been very obliging in communicating them to me.—It is much to be desired, that like accounts were kept in every town and parish. It would be extremely agreeable to learn from them the different rates of human mortality in different places, and the number of people and progress of population in the kingdom. The trouble of keeping them would be trifling; but the instruction derived from them (*a*), would be very important.—I have already proposed one improvement of such accounts. I will add, that they would be still more useful, did they give the ages of the dead after 10, within periods of *five,* instead of *ten* years.—During every period, so short as *five* years, the decrements

(*a*) See Essay I. p. 210, 211.

of life may, in conftructing Tables, be fafely taken to be *uniform*. But this cannot be equally depended on, in periods fo long as ten years.

There is yet another improvement of thefe accounts, which I fhall take this opportunity to mention. They fhould contain not only a lift of the diftempers of which all die, like that in the *London* bills; but they fhould fpecify particularly the numbers dying of thefe diftempers, in the feveral divifions of life (*a*). Accurate regifters of mortality kept in this manner, in all parts of the kingdom; and compared with records of the feafons, and of the weather, and with the particular circumftances which difcriminate different fituations, might contribute, more than can be eafily imagined, to the increafe of *phyfical* knowledge.—But to proceed no farther in thefe Obfervations; I fhall now beg leave to fhut up this whole work with the following general reflection.

I have reprefented particularly, the great difference between the probabilities of human life in towns and in country parifhes; and from the facts I have recited, it appears, that the further we go from the artificial and ir-

---

(*a*) Since the former editions of this work, bills, on an improved plan of this kind, have been actually eftablifhed at *Manchefter* and *Chefter*.

regular modes of living in great towns, the fewer of mankind die in the *firſt* ſtages of life, and the more in its *laſt* ſtages. The lower animals (except ſuch (*a*) as have been taken under human management) ſeem in general to enjoy the full period of exiſtence allotted them, and to die chiefly of old age: And were any obſervations to be made among *ſavages,* perhaps the ſame would be found to be true of them.—DEATH is an evil to which the order of Providence has ſubjected every inhabitant of this earth; but to man it has been rendered unſpeakably more an evil than it was deſigned to be. The greateſt part of that black catalogue of diſeaſes which ravage human life, is the off-ſpring of the tenderneſs, the luxury, and the corruptions introduced by the vices and falſe refinements of

---

(*a*) Calves are the only animals taken under our peculiar care immediately after birth; and, in conſequence of then adminiſtring to them the ſame ſort of phyſic that is given to *infants,* and treating them in other reſpects in the ſame manner, it is probable, that more of them die ſoon after being born, than of *all* the other ſpecies of animals, which we ſee in the ſame circumſtances. See the *Comparative View of the State and Faculties of Man with thoſe of the Animal World,* p. 23.—It is, indeed, melancholy to think of the havock made among the human ſpecies by the unnatural *cuſtoms* as well as the *vices,* which prevail in poliſhed ſocieties. I have no doubt, but that the cuſtom, in particular, of committing infants, as ſoon as born, to the care of *foſter-mothers,* deſtroys more lives than the ſword, famine, and peſtilence put together.

civil society (*a*). That delicacy which is injured by every breath of air, and that rottenness of constitution which is the effect of indolence, intemperance and debauchery, were never intended by the Author of Nature; and it is impossible, that they should not lay the foundation of numberless sufferings, and terminate in premature and miserable deaths.—Let us then value more the simplicity and innocence of a life agreeable to nature; and learn to consider nothing as savageness but malevolence, ignorance, and wickedness. The order of nature is wise and kind. In a conformity to it consists health and long life; grace, honour, virtue and joy. But nature turned out of its way will always punish. *The wicked shall not live out half their days.* Criminal excesses embitter and cut short our *present existence*; and the highest authority has taught us to expect, that they will not only kill the *body*, but the *soul*; and deprive of an EVERLASTING EXISTENCE.

(*a*) The ingenious and excellent writer quoted in the last note, observes, that the whole class of diseases which arise from catching cold, are found only among the civilized part of mankind, p. 51.—And, concerning that loss of all our higher powers which so often attends the decline of life, and which is so humiliating to human pride; he observes, that it exhibits a scene singular in nature, and that there is the greatest reason to believe, that it proceeds from adventitious causes, and would not take place among us if we led natural lives, p. 62.

APPEN-

APPENDIX.

Note (A). See Question III. Page 11.

LET E be any given expectation of life; and $\frac{4E-x}{4E} \times px$ will be the number of persons alive at the end of x years, arising from p persons left annually as widows, (or added annually to a town or society) at the age whose *expectation* is E. The *maximum*, therefore, is always pE—. In Mr. *De Moivre's Hypothesis*, E is always $\frac{1}{2}$ the difference between the given age and 86. See the note, page 2, and the latter end of the note in page 37. See likewise the beginning of the First Essay, and note (L) in this Appendix, where the investigation of this rule will be given.

It will not be amiss to give the following example of the application of this rule.

At the time of the commencement of the scheme, among the ministers and professors in SCOTLAND, for making provision for their widows, it was necessary, that a calculation should be made of the number of widows that would be upon the scheme at the end of every year, till they came to a *maximum*, on the supposition that, (agreeably to what particular enquiry had shewn to have happened for many preceding years,) 20 new widows would be left every year (*a*). In order to make

---

(*a*) For the last 27 years; that is, from the commencement of the scheme to the present time, this number has been $19\frac{1}{4}$, as mentioned, p. 275.

this calculation, let 4 of the 20 widows be supposed to be under 32 years of age when left; and let 28 be supposed their mean age. Let the same number be left between 32 and 39, and let 35 be their mean age; between 39 and 47, and 43 their mean age; between 47 and 57, and 52 their mean age; between 57 and the extremity of life, and 63 their mean age. The number in life together, to which, in 10 years, 4 widows left annually at the age of 28 will grow, is, by the rule, (E being 29) $\frac{116-10}{116} \times 40$, or 36.55.——The number alive at the end of 20 years, will be $\frac{116-20}{116} \times 80$, or 66.2. At the end of 30 years, the number alive will be 89; of 40 years, 104.82: of 58 years 116——These numbers, found in the same way, for the 2d class, (E being 25.5,) at the end of 10, 20, 30, 40, and 51 years, will be 36.7—64.31—84.7—97.25—102——For the 3d class, (E being 21.5) at the end of 10, 20, 30, 40, and 43 years, 35.34—61.4—78.13—85.6—86——For the 4th class, (E being 17) at the end of 10, 20, 30, and 34 years, 34.11—56.47—67—68——For the 5th class, (E being 11.5) at the end of 10, 20, and 23 years, 31.3—45.2—46——The whole number, therefore, consisting of all the classes, will come to a *maximum* nearly in 58 years; and the totals in life, at the end of 10, 20, 30, 40, 50, and 58 years, will be 173.37—293.58—364.83—401.67—418.

These determinations suppose none to marry. In 10 years, from 1757 to 1767, I have been informed, that but 9 widows married. Let us then suppose, that one widow of the first class marries every year; and let all that marry, be supposed to continue, one with another, 5 years in widowhood

hood before they marry. On thefe fuppofitions, the foregoing totals will, at the end of the fame periods of years, be 169.23 — 282 — 347.5 — 380.47 — 394.

Thefe calculations are made from Mr. *De Moivre*'s Hypothefis. Had they been made exactly from Dr. *Halley*'s Table, or any other of the Tables I have given at the end of this work, except the *London* one, the refults would have been very nearly the fame.

Twenty-feven years have now elapfed fince the commencement of this fcheme; and the number of widows living every year have, in fact, corresponded to the laft numbers I have given, as nearly as could be expected.

Note (B).   Question VI. Page 21.

LET r signify the sum of 1 l. and its interest, for one year. The value of a life, whose complement is n, being (by Mr. *De Moivre* on *Annuities*, 4th edition, page 14. and p. 100.) $\frac{n-1}{nr} + \frac{n-2}{nr^2} + \frac{n-3}{nr^3} + \frac{n-4}{nr^4}$, &c. the present value of the remainder of it after *two* years must be $\frac{n-3}{nr^3} + \frac{n-4}{nr^4}$, &c. which is equal to $\frac{1}{r^2} \times \frac{n-2}{n} \times \overline{\frac{n-3}{n-2r} + \frac{n-4}{n-2r^2} + \frac{n-5}{n-2r^3}}$, &c.

Now $\frac{1}{r^2}$ is the present value of 1 l. due at the end of two years. $\frac{n-2}{n}$ is the probability that a life, whose complement is n, shall continue two years, and $\frac{n-3}{n-2r} + \frac{n-4}{n-2r^2} + \frac{n-5}{n-2r^3}$, &c. is the value of a life two years older than the life whose complement is n. And, therefore, (since any number of years less than n may be substituted for two years) the first rule given in this Question is right.

The same process, applied to joint lives, will demonstrate what is said in the *Scholium*.

Note,

APPENDIX.

Note (C). See Question VII. Page 22.

LET the complements of any two assigned lives be n and m. The present value of the first possible payment of an annuity to be enjoyed by the life whose complement is n, provided *both* lives continue 7 years, and the life, whose complement is n, survives the other *after* that term, is the probability, that the life of the expectant shall continue 8 years, and the other life 7 years and then fail in the 8th year, multiplied by $\frac{1}{r^8}$, or by 1 *l.* discounted for 8 years.—The probability that the life of the *expectant* shall continue 8 years is $\frac{n-8}{n}$. The probability that the *other* life shall continue 7 years is $\frac{m-7}{m}$. The probability that it shall continue 7 years, and fail in the 8th year, is $\frac{m-7}{m} \times 1 - \frac{m-8}{m-7} = \frac{1}{m}$. The probability, therefore, that the life of the *expectant* shall continue 8 years, and the other life continue 7 years and fail in the 8th, is $\frac{n-8}{n} \times \frac{1}{m}$; and the present value of the first possible payment of the annuity supposed, is $\frac{n-8}{nr^8} \times \frac{1}{m}$. See *The Doctrine of Annuities*, by Mr. *Simpson*, p. 6—15, or his *Select Exercises*, p. 315, &c.——In like manner, the present value of the 2d payment, at the end of the 9th year, may be found to be $\frac{n-9}{nr^9} \times \frac{m-7}{m} \times \overline{1 - \frac{m-9}{m-7}}$, or $\frac{n-9}{nr^9} \times \frac{2}{m}$. and the present value of all the possible payments,

$\frac{1}{r^7} \times \overline{\frac{n-8}{nr} \times \frac{1}{m} + \frac{n-9}{nr^2} \times \frac{2}{m} + \frac{n-10}{nr^3} \times \frac{3}{m}}$, &c.

But this series is equal to $\frac{1}{r^7} \times \frac{n-7}{n} \times \frac{m-7}{m} \times$

$\overline{\frac{n-8}{n-7r} \times \frac{1}{m-7} + \frac{n-9}{n-7r^2} \times \frac{2}{m-7} + \frac{n-10}{n-7r^3} \times}$

$\frac{3}{m-7}$, &c. Now $\frac{n-8}{n-7r} \times \frac{1}{m-7} + \frac{n-9}{n-7r^2} \times \frac{2}{m-7}$, &c. is the value of an annuity for a life seven years older than the expectant, after another life seven years older than the life whose complement is m. $\frac{n-7}{n} \times \frac{m-7}{m}$ is the probability that both the assigned lives shall continue 7 years. And $\frac{1}{r^7}$ is the value of 1 *l.* due at the end of 7 years. The rule, therefore, given for solving this question, is right.

This demonstration, as well as that in the last note, is, for the sake of more ease and clearness, applied to the hypothesis of an equal decrement of life. It does not, however, depend upon it, but may be applied to any table of observations.

Note,

APPENDIX.

Note (D). Question IX. Page 29.

LET the complement of any two assigned lives be n and m, and the given term be *seven* years, as in note (C). The probability that the former life (supposed to be the life in expectation) shall last 8 years, is, by Mr. *De Moivre*'s Hypothesis, $\frac{n-8}{n}$; and the probability that the latter life shall fail in 8 years, is $\frac{8}{m}$; and the first payment of the annuity mentioned in this question, depends on the happening of *both* these events, the probability of which is $\frac{n-8}{n} \times \frac{8}{m}$.

The present value, therefore, of the first possible payment of the annuity is $\frac{n-8}{nr^8} \times \frac{8}{m}$. —— In like manner; the present value of the *second* possible payment is $\frac{n-9}{nr^9} \times \frac{9}{m}$; and of all the payments, $\frac{n-8}{nr^8} \times \frac{8}{m} + \frac{n-9}{nr^9} \times \frac{9}{m} + \frac{n-10}{nr^{10}} \times \frac{10}{m}$, &c. But $\frac{n-8}{nr^8} \times \frac{8}{m} = \frac{n-8}{nr^8} \times \frac{1}{m} + \frac{n-8}{nr^8} \times \frac{7}{m}$; and $\frac{n-9}{nr^9} \times \frac{9}{m} = \frac{n-9}{nr^9} \times \frac{2}{m} + \frac{n-9}{nr^9} \times \frac{7}{m}$. The foregoing series, therefore, is equal to the two series's $\frac{1}{r^7} \times \frac{n-8}{nr} \times \frac{1}{m} + \frac{n-9}{nr^2} \times \frac{2}{m} + \frac{n-10}{nr^3} \times \frac{3}{m}$, &c. and $\frac{1}{r^7}$

$$\frac{1}{r^7} \times \overline{\frac{n-8}{nr} \times \frac{7}{m} + \frac{n-9}{nr^2} \times \frac{7}{m} + \frac{n-10}{nr^3} \times \frac{7}{m}}, \&c. \text{ or}$$

to $\frac{1}{r^7} \times \frac{n-7}{n} \times \frac{m-7}{m} \times \overline{\frac{n-8}{n-7r} \times \frac{1}{m-7} + \frac{n-9}{n-7r^2}} \times$

$\overline{\frac{2}{m-7} + \frac{n-10}{n-7r^3} \times \frac{3}{m-7}}$, &c. $+ \frac{1}{r^7} \times \frac{7}{m} \times \frac{n-7}{n} \times$

$\overline{\frac{n-8}{n-7r} + \frac{n-9}{n-7r^2} + \frac{n-10}{n-7r^3}}$, &c. which is the very rule given for solving this queſtion, as will appear from notes (B) and (C).

Note,

Note (E). See the Scholium to Quest. X.

ACCORDING to the calculations, the time in which the first yearly payment of a reversionary *annuity* becomes due, is the end of the year in which the event happens that entitles to it, however little or much of the year may then happen to be unelapsed. And this, likewise, is the time when a reversionary *sum* becomes due. Those who know how the calculations of the values of reversions are instituted, must know this. But an annuity, the first payment of which is to be made at the same time with another payment of a sum in hand, sufficient to buy an equal annuity, is worth one year's purchase more than that sum. For instance. Reckoning interest at 4 per cent. and r being $1l.$ increased by its interest for a year, or 1.04, $\frac{1}{r} + \frac{1}{r^2} + \frac{1}{r^3}$, &c. $= 25 l.$ is the present value of an estate of $1 l.$ per annum for ever. That is, it is the value of it, supposing the first rent of it is to be paid a year hence.——If the first rent is to be received immediately, or at the same time with another payment of $25 l.$ it is worth one year's purchase more, or equivalent to $26 l.$——I have not found, that any of the writers on annuities and reversions, have attended to this observation. It suggests a correction necessary to be applied to the common solutions of several important problems: particularly to the 21st and 22d in Mr. *Simpson's Treatise on Annuities*, and the 26th, 27th, 32d, 33d, and 40th problems in his *Select Exercises*; and to all other problems of the same kind in other writers. There can

can be no great occasion for being more explicit. It will not, however, be amiss to add the following demonstration. —— $\frac{1}{n}$ is the present probability that a life whose complement is n will fail in any one assignable year of its duration. $S \times \overline{\frac{1}{nr} + \frac{1}{nr^2} + \frac{1}{nr^3}}$, &c. ($n$), or the present value of 1 *l.* per annum for n years, multiplied by $\frac{S}{n}$, is the present value of the sum or legacy denoted by S, payable at the failure of the given life. Therefore, (n being 56; the life 30; interest 4 per cent. $r = 1.04$; the sum 25 *l.*) the value of the expectation, by Mr. *De Moivre*'s hypothesis, is 9.919.

Further. The value of 1 *l.* to be received at the end of a year, provided the life whose complement is n fails, is the probability of the failure of the life multiplied by 1 *l.* discounted for a year, or $1 - \overline{\frac{n-1}{n}} \times \frac{1}{r}$. In like manner; the value of 1 *l.* to be received at the end of two years, if the same life fails in 2 years, is $1 - \overline{\frac{n-2}{n}} \times \frac{1}{r^2}$. And, therefore, the value of all the *possible* payments of an estate or annuity of 1 *l.* for ever, to be entered upon after the given life, is $1 - \overline{\frac{n-1}{n}} \times \frac{1}{r} + 1 - \overline{\frac{n-2}{n}} \times \frac{1}{r^2} + 1 - \overline{\frac{n-3}{n}} \times \frac{1}{r^3}$, &c. ($n$) $+ \frac{1}{r^{n+1}} + \frac{1}{r^{n+2}}$

$\frac{1}{r^{n+2}}$, &c. or $\frac{1}{r} + \frac{1}{r^2} + \frac{1}{r^3}$, &c. $- \overline{\frac{n-1}{nr} + \frac{n-2}{nr^2} +}$ $\overline{\frac{n-3}{nr^3}}$, &c. that is, the value of the life subtracted from the perpetuity; or, in this example, *l*. 14.684, (the value of a life at 30) subtracted from 25; that is, *l*. 10.316. But 10.316 is to 9.919, in the same ratio with 104 to 100, or 26 to 25, agreeably to the rule in the *Scholium*.

Note (F). Question XIII. Page 44.

WHEN I here call 48 the mean age of all married men, and 40 the mean age of married women, I do not intend to suppose, that there are as many married persons who exceed these ages, as there are who fall short of them. It is likely that the latter are most numerous; and it is necessary that this should be the case, to render the supposition I make just.—If all marriages commenced at 33 for the man, and 25 for the woman, one half of them would be dissolved by the time the men were 50, and the women 42; for (by the *Hypothesis*, and also nearly by the *Breslaw*, *Norwich*, and *Northampton* tables) there is an equal chance for the joint continuance of two lives, whose ages are 25 and 33, *seventeen* years. Forty-two and fifty then would be properly the mean ages at which widowhood would commence; meaning by these " the ages on each side of which equal numbers are left widows and widowers."——But, tho' in this case half the marriages of every year would be dissolved in 17 years, they would not be *all* dissolved in twice that time. So far would this be from happening, that about a 7th part would continue beyond twice 17 years; nor would it be *certain*, that they would be all dissolved till near the extremity of the possible extent of life. Tho', therefore, an equal number of marriages would be dissolved, or an equal number of widows and widowers left *before* 50 and 42, and *afterwards*, yet the ages of the latter would, one with another, much more exceed 50 and 42, than the ages of the former (that is, of the widows and widowers left

APPENDIX.

before 50 and 42) would fall short of them. And the number of marriages also in the world, among persons of greater ages than these, would be much fewer than among persons of lesser ages.—In other words: the period, at which the marriages that have been contracted are half dissolved, is not the period at which the number of marriages constantly existing is equally divided, but this period falls some years sooner; and the period I have in view, falls in that part of the interval between these two periods, where the greater ages of the marriages on one side, are just enough to compensate (in such a calculation as that I have given) their deficiencies in number, compared with the number of marriages on the other side.

In short. Suppose 35 marriages every year, between persons 33 and 25 (*a*). In 12 years there would be half as many in the world, as could possibly arise from such a number of yearly weddings. In 17 years, half every set would be extinct. The *expectation* of every marriage would be 19 years, by prob. 21 of Mr. *De Moivre's Treatise on Annuities*, or by the note p. 305: That is, taking them all together, they would exist just as long as an equal number of *single* persons, supposed to be sure of living just 19 years, and no more: or, as long as an equal number of single persons, all 48 years of age, supposed to be subject to the common laws of mortality. One with another, then, they will be all extinct in 19 years: the marriages which continue beyond this term, tho' fewer in number, enjoying among them just as

(*a*) In the *Pais de Vaud, Switzerland*, the mean age at which women marry, is nearly the very age here mentioned: But it will be shewn in the *Supplement*, that the expectation of marriage there, is no less than 23 years and $\frac{1}{2}$; so much higher are the probabilities of life in the *country* than in *towns*, or than they ought to be according to Mr. *De Moivre's Hypothesis*. See p. 268.

much *more* duration, as those that fall short of it enjoy *less*. *Widows*, then, at a medium, will commence widowhood at 44 (that is, 25 increased by 19) years of age, and *widowers* at 52. The values, therefore, of the lives of the *former*, when they commence widowhood, will, one with another, be the same with the value of a life at 44; or, (reckoning interest at 4 per cent.) 12.5 years purchase, in one present payment, (the annuity to begin at the end of a year); and their *expectation* of life will be 21 years, or half the difference between 44 and 86. The value of the lives of the *latter* will be 10.92, and their *expectation* 17 years.—The whole number of marriages constantly existing, which would result from 35 supposed to commence annually, would be 19×35, or 665; and 53 years (the difference between 33 and 86) would be the time in which they would increase to this number—The chance of survivorship would be the odds of 69 to 53, by prob. 18th, Mr. *De Moivre on Annuities*; that is, in 53 years, 35 relicts of these marriages would be left every year, and the number of *widows* would be to the number of *widowers*, as 69 to 53; or 19.8 *widows* would be left annually, and 15.2 *widowers*. The *maximum* of widows in life together, if none married, would be 21×19.8, or 416; and they would increase to this number in 114 years (or 61 years after the number of marriages had attained to a *maximum*)——The *maximum* of *widowers* would be 15.2×17, or 258; and they would increase to this number in 106 years.

An easy method may be hence deduced of solving the question which occasions this note——If the number of the members of the establishment I have supposed, is 665, and the mean ages at which marriage may be deemed to commence are 25 and 33, 19.8 widows will (it has just appeared) be be left every year; and the values of their lives, when they commence widowhood, will be, one with another, $12\frac{1}{2}$ years purchase. An annuity of 20 *l*. will, therefore, be worth, to each widow, 250 *l*. and 19.8 such annuities must be worth 4950 *l*. which, consequently, is the annual income necessary for the support of the establishment, the first payment to be received immediately: or *l*. 7.44 from each of the 665 members; which answers nearly to the determination in the note page 44.

In the last Essay, p. 275, it has been shewn, that observations determine the chance of survivorship in favour of the wife in marriage, to be really so great as 3 to 2; and in some circumstances greater. I have also there observed, that in order to account for this, from the difference of age between men and their wives, this difference must be at least 12 years, and the mean ages of all who marry annually, must be supposed to be about 23 and 35. In this case, 19, as before, will nearly be the *expectation* of all marriages. The mean age at which widows and widowers will commence such will be 42 and 54. The number of annual marriages necessary to keep up 665 marriages constantly existing, will be 35. The number of widows left annually, by such a number of marriages, will be 21; and the values of their lives, at the time they commence widowhood, will be 12.85 years purchase by Table VI: and therefore, the whole annual income necessary for the support of the supposed establishment, will be 539 *l*. or an annual payment, beginning immediately, of *l*.8.11 from each member.—The number of widows on such an establishment will, in 63 years, grow, if none marry, to 462; and the number of widowers to 224.—It may be depended on, that all this would happen as far as Dr. *Halley*'s Table, or the Tables for *Norwich* and *Northampton*, exhibit the true state of human mortality.

Among

Among the ministers and professors in Scotland, the number of married men being 667, or nearly that here mentioned, the number of annual weddings has, for many years, been at an average 31, and the number of widows left annually 19.2; and, therefore, the chance of survivorship in favour of the wife, as 19.2 to 11.8, or 5 to 3. See Essay IV. p. 274. This is not more different from the results I have given, than might have been expected; and the chief reason of the difference is, that the *expectations* of *single* and *joint* lives among the ministers and their wives in Scotland, are greater than those given by Dr. *Halley*'s, and the other tables of observation——These tables give the expectations of lives as they are among the bulk of mankind in moderate towns. The expectations of lives among the better sort of men, living mostly in country villages and parishes, are much greater. The fact is, that among the ministers in *Scotland*, the expectation of a *single* life, at the age of 27, is three years and an half greater; and, of *joint* lives, about two years and a half greater, than the same expectations by Dr. *Halley*'s Table. Ibid. page 269.

I cannot help just mentioning another remark here.———It may be observed, that supposing no second marriages, and, at the same time, that the odds for the woman's surviving in marriage is 3 to 2, the number of *widows* in the world would be *double* the number of *widowers*. But it has been found, in fact, that the number of widows is five times the number of widowers. How this is to be accounted for, I have shewn in the Essay just referred to, page 276.

APPENDIX.

Note (G). Question XIV. Page 48.

LET r be $1l.$ increased by its interest for one year; t the given time or number of years for which the assurance is to be made; $a, b, c, \&c.$ the *probabilities* taken out of a table of observations, that the person whose age is given shall live 1, 2, 3, &c. years; and P the probability that he shall live t years. Then $\overline{\frac{1-a}{r} + \frac{1-b}{r^2} + \frac{1-c}{r^3}, \&c. (t-1)} + \frac{1-P}{r^t} + \frac{1-P}{r^{t+1}} + \frac{1-P}{r^{t+2}}, \&c. = \frac{1}{r} + \frac{1}{r^2} + \frac{1}{r^3}, \&c. (t) - \overline{\frac{a}{r} + \frac{b}{r^2} + \frac{c}{r^3}, \&c. (t-1)} + \frac{P}{r^t} + \frac{1-P}{r^t} \times \frac{1}{r} + \frac{1}{r^2} + \frac{1}{r^3}, \&c.$ will be the exact value of an annuity to be entered upon at the failure of the given life, provided it happens in t years. And the rule is nothing but this value expressed in words. In a similar manner may be demonstrated the other rule for finding the values of assurances for a given time, on two joint lives, or the longest of two lives.

Note (H).   Question XV. Page 56.

LET r signify as before; S the given sum to be assured; t the given time; N and n the number of the living in the table of observations, at the age of A and B respectively; A, B, C, &c. and a, b, c, &c. the number of the living in the table, at the end of 1, 2, 3, &c. years from the ages of A and B; $\overline{D}, \overset{\shortmid}{D}, \overset{\shortparallel}{D}, \overset{\shortmid\shortmid\shortmid}{D}$, &c. and $\overline{d}, \overset{\shortmid}{d}, \overset{\shortparallel}{d}, \overset{\shortmid\shortmid\shortmid}{d}$, &c. the decrements of life in the table, at the end of 1, 2, 3, &c. years from the same ages. Then, by reasoning in the same manner with Mr. *Simpson*, in p. 316, &c. *Select Exercises*, it will appear that S $\times$ $\overline{\frac{A \times d}{Nnr}} + \overline{\frac{B \times \overset{\shortmid}{d}}{Nnr^2}} + \overline{\frac{C \times \overset{\shortparallel}{d}}{Nnr^3}}$, &c. $(t) +$ S $\times$ $\overline{\frac{\overline{D}d}{2Nnr}} + \overline{\frac{\overset{\shortmid}{D}\overset{\shortmid}{d}}{2Nnr^2}} + \overline{\frac{\overset{\shortparallel}{D}\overset{\shortparallel}{d}}{2Nnr^3}}$, &c. $(t) = \frac{S}{n} \times \overline{\frac{Ad}{Nr}} + \overline{\frac{B\overset{\shortmid}{d}}{Nr^2}} + \overline{\frac{C\overset{\shortparallel}{d}}{Nr^3}}$, &c. $(t) + \frac{S}{2N} \times \overline{\frac{\overline{D}d}{nr}} + \overline{\frac{\overset{\shortparallel}{D}\overset{\shortparallel}{d}}{nr^2}}$, &c. (t). This is the exact answer to Question XV. and the rule is as near an approximation to it as there is reason to desire.

In the same manner, retaining all the same symbols, it may be found, that the answer to Question XVI. is

S $\times$ $\overline{\frac{\overline{D}d}{2Nnr}} + \overline{\frac{\overset{\shortmid}{D}\overset{\shortmid}{d}}{Nnr^2}} + \overline{\frac{\overline{D+\overset{\shortmid}{D}} \times \overset{\shortparallel}{d}}{Nnr^3}} + \overline{\frac{\overline{D+\overset{\shortmid}{D}+\overset{\shortparallel}{D}} \times \overset{\shortmid\shortmid\shortmid}{d}}{Nnr^4}}$

(t), &c. $+$ S $\times$ $\overline{\frac{\overset{\shortmid}{D}\overset{\shortmid}{d}}{2Nnr^2}} + \overline{\frac{\overset{\shortparallel}{D}\overset{\shortparallel}{d}}{2Nnr^3}} + \overline{\frac{\overset{\shortmid\shortmid\shortmid}{D}\overset{\shortmid\shortmid\shortmid}{d}}{2Nnr^4}}$, &c.

$(t-1)$

$$(t-1) = \frac{S}{nr} \times \frac{\overline{Dd}}{Nr} + \frac{\overline{D+D} \times d}{Nr^2} + \frac{\overline{D+D+D} \times d}{Nr^3},$$

&c. $(t-1) + \frac{S}{2N} \times \frac{\overline{Dd}}{nr} + \frac{\overline{Dd}}{\frac{\text{\tiny I I}}{nr^2}} + \frac{\overline{Dd}}{\frac{\text{\tiny II II}}{nr^3}}$, &c. (t).

But $\frac{D}{Nr} + \frac{\overline{D+D}}{Nr^2} + \frac{\overline{D+D+D}}{Nr^3}$, &c. $(t-1)$ is the same with the excess of the value of an annuity *certain* for a number of years less by one year than the given term, above the value of an annuity on the life of A, for the same number of years; from whence the reason of the rule for solving this question may be easily discovered.

Note

Note (I). Page 118, &c.

LET t be any given term of years; p the value of $1 l.$ due at the end of the given term; A the value of an annuity certain for the same term; n the *complement* of a given life; G the value for the given term, of two joint lives, both equal to the given life; (to be found by Quest. VI.) P the perpetuity; r, $1 l.$ increased by its interest for one year.

Then $\overline{A-G} \times n+t \times p \times P - A \times P \times r$ will be the present value of $1 l.$ $2 l.$ $3 l.$ &c. (t) payable at the end of $1, 2, 3,$ &c. (t) years; but subject to failure when the given life fails.

If such a course of payment is to begin immediately, and to be made at the beginning of every year, till $t+1$ payments are made in t years; add to the preceding value, the value increased by unity of an annuity on the given life for t years, found by Question VI. and the *sum* will be the value sought. And this value divided by the present value of what may happen to remain of the given life after t years, found by Question VI. will give the *standing annuity* to which such a series of increasing annual payments, beginning immediately, will entitle, for the remainder of the given life after t years.

With the assistance of this theorem, all that is said in p. 117, &c. may be investigated. It would be too tedious to enter into a more minute account.

Note (K). Page 149.

LET d signify the *difference* between the *complements* of the youngest and oldest life in the body of Annuitants, here described, at the time they enter; let S signify the sum of these *complements*; n any given number of years not greater than $\frac{S}{2} - \frac{d}{2}$; and x the ratio of the whole number of Annuitants to $\frac{S \times d}{2}$. Then

$x \times d$ will be the number that will die the 1st year;

$x \times \overline{d + \frac{2d}{S}}$, the number that will die the 2d year;

$x \times \overline{d + \frac{4d}{S} + \frac{4d}{S^2}}$, 3d year;

$x \times \overline{d + \frac{6d}{S} + \frac{8d}{S^2} + \frac{8d}{S^3}}$, 4th year;

$x \times \overline{d + \frac{8d}{S} + \frac{12d}{S^2} + \frac{16d}{S^3} + \frac{16d}{S^4}}$, 5th year;

and $x \times \overline{nd + \overline{n^2 - n} \times \frac{d}{S} + \overline{n-2} + \overline{n-2}|^2 \times \frac{2d}{S^2} + \overline{n-3}}$
$+ \overline{n-3}|^2 \times \frac{4d}{S^3} + \overline{n-4} + \overline{n-4}|^2 \times \frac{8d}{S^4}$, &c. (n)

will be the whole number dying in n years. When n is greater than $\frac{S}{2} - \frac{d}{2}$, this series is greater than the whole number dying in n years; but in all other cases it gives this number exactly, supposing the probabilities of life to decrease uniformly.——

In

In the present instance, the youngest life being 30, and the oldest 60, the two complements are 56 and 26. $S = 82$. $d = 30$. $\frac{Sd}{2} = 1230$. And therefore $x = \frac{33.333}{1230} = 27.1$. Take $n = 30$ years, and the foregoing series will be $27.1 \times 900 + 318.2 + 7.242 + .164 = 33.214$, which is a little greater than the whole number dying in 30 years; but at the same time less than the whole number of Annuitants.

Note

APPENDIX. 305

Note (L). See Essay I. Page 170, 171, 173.

THE *sum* of the probabilities that any given lives will attain to the end of the 1st, 2d, 3d, &c. *years* from the present time to the utmost extremity of life (for instance, $\frac{45}{46} + \frac{44}{46} + \frac{43}{46}$, &c. to $\frac{1}{46} = 22\frac{1}{2}$ for lives of 40, by the *hypothesis*) may be called their *expectation*, or the number of payments due to them, as *yearly annuitants*. The sum of the probabilities that they will attain to the end of the 1st, 2d, 3d, &c. *half years*, (or, in the particular case specified, $\frac{91}{92} + \frac{90}{92} + \frac{89}{92} + \frac{88}{92}$, &c. $= \frac{91}{4}$ *half years*, or $22\frac{3}{4}$ *years*) is their expectation as *half yearly annuitants*. And the sums just mentioned of the probabilities of their attaining to the end of the 1st, 2d, 3d, &c. *moments* (equal in the same particular case to 23 years) is properly their *expectation of life*, or their *expectation* as annuitants secured by land.

Mr. *De Moivre* has omitted the demonstrations of the rules he has given for finding the *expectations* of lives, and only intimated in general, that he discovered them by a calculation deduced from the method of fluxions. See his *Treatise on Annuities*, page 66. It will, perhaps, be agreeable to some to see how easily they are deduced in this method, upon the hypothesis of an equal decrement of life.

Let $\dot{x}$ stand for a moment of time, and n the *complement* of any assigned life. Then $\frac{n-\dot{x}}{n}$, $\frac{n-2\dot{x}}{n}$, $\frac{n-3\dot{x}}{n}$, &c. will be the *present* probabilities of its

X con-

continuing to the end of the 1ft, 2d, 3d, &c. moments; and $\frac{n-x}{n}$ the probability of its continuing to the end of x time. $\frac{n-x}{n} \times \dot{x}$ will therefore be the *fluxion* of the sum of the probabilities, or of an *area* representing this sum, whose *ordinates* are $\frac{n-x}{n}$, and *axis* x.—The *fluent* of this expression, or $x - \frac{x^2}{2n}$, is the sum itself for the time x; and this, when $x = n$, becomes $\frac{1}{2}n$, and gives the *expectation* of the assigned life, or the sum of all the probabilities just mentioned, for its whole possible duration.—In like manner: since $\frac{\overline{n-x}^2}{n^2}$ is the probability that two equal joint lives will continue x time, $\frac{\overline{n-x}^2}{n^2} \times \dot{x}$ will be the *fluxion* of the sum of the probabilities. The *fluent* is $x - \frac{x^2}{n} + \frac{x^3}{3n^2}$, which, when $n = x$, is $\frac{n}{3}$, or the expectation of two equal joint lives.—Again: since $\frac{n-x}{n} \times \frac{2x}{n}$ is the probability that there will be a survivor of two equal joint lives at the end of x time, $\frac{n-x}{n} \times \frac{2x}{n} \times \dot{x}$ will be the *fluxion* of the sum of the probabilities; and the *fluent*, or $\frac{x^2}{n} - \frac{2x^3}{3n^2}$ is (when $x = n$) $\frac{1}{3}n$, or the *expectation* of survivorship between two equal lives; which, therefore, appears to be equal to the

expecta-

APPENDIX.

expectation of their joint continuance. The expectation of two *unequal* joint lives, found in the same way, is $\frac{m}{2} - \frac{m^2}{6n}$, m being the *complement* of the oldest life, and n the *complement* of the youngest. The whole expectation of survivorship is $\frac{n}{2} - \frac{m}{2} + \frac{m^2}{3n}$. And the expectation of survivorship of the oldest will be to the expectation of survivorship of of the youngest, as $\frac{m^2}{6n}$ to $\frac{n}{2} - \frac{m}{2} + \frac{m^2}{6n}$. It is easy to apply this investigation to any number of joint lives, and to all cases of survivorship.

It may be observed, concerning the first of the fluents here given, that it expresses not only the expectation of a given life for the time x, and therefore its whole expectation when $x = n$, but likewise, the number of persons alive, to which one person added annually to a society, at a given age, will increase in x time.——Thus: Suppose one annuitant, whose age is 28, (and whose *complement* of life, therefore, is 58, or *expectation* of life 29) to come upon a society every year; the number of annuitants alive, deduced from hence, will, in x years, be $x - \frac{x^2}{4 \times 29}$, or $\frac{4 \times 29 - x^2}{4 \times 29} \times x$; and, therefore, the number of annuitants alive, deduced in the same time from p annuitants left annually at the same age, will be $\frac{4 \times 29 - x^2}{4 \times 29} \times px$.——In like manner, the 2d fluent, or $\frac{x^3}{3n^2} - \frac{x^3}{n} + x$, gives the number of marriages in being together, that will, in x years, grow out of *one* yearly marriage, between persons of *equal* ages, whose complement of life is n. If they are of *unequal* ages, and the complement of the oldest life is m, and of the youngest n, this number will be $\frac{x^3}{3nm} - \frac{\overline{n+m} \times x^2}{2nm} + x$. And if the number of years is required, in which any given number of yearly marriages, between men and women at given ages, will increase so far as to be in any given proportion to the greatest number that can possibly grow out of such marriages, this expression must be made equal to the *expectation* of the joint lives, or of each marriage, multiplied by the fraction expressing the given proportion; and the root of the equation will be the answer. Thus: it may be found, that one marriage every year, between persons 33 and 25 years of age, would in 10 years increase to 8.35; in 15 years, to 11.38; and in 53 years, to 19, or their greatest possible number; and, consequently, that 35 such yearly marriages would, in 10 years, increase to 292; in 15 years, to 398; and in 53 years, to 665.——And if it is enquired in what number of years 35 such yearly marriages would increase to half the number in being together, possible to be derived from them, the value of x, in the cubic equation $\frac{x^3}{3nm} - \frac{\overline{n+m} \times x^2}{2nm} + x = \frac{m}{2} - \frac{m}{6n} \times \frac{1}{2}$, must be found; which, in the present instance, is nearly 12.

I have, in some parts of this work, had occasion to make such deductions as these. See note (A), p. 283; and note (F), p. 294; and Questions III. and XIII.

Note

Note (M). Essay II. Page 231.

LET r signify $1l.$ increased by its interest for one year.

V the PERPETUITY.

n the difference between the age of the youngest life, and 86; or its *complement.*

m the complement of the oldest life.

P the value (in Table II.) of an annuity certain for m years.

And the exact value of any two given joint lives, according to the hypothesis of an equal decrement of life, will be $V - \frac{V+1}{n} \times \overline{n-m-2v-1} \times \frac{P}{m} + 2v.$ Example:

Let the ages be 27 and 38; and the rate of interest 4 *per cent.* Then $n = 59.$ $m = 48.$ $V = 25.$ $P = 21.195.$ $\overline{n - m - 2v - 1} = -40.$ $\overline{n-m-2v-1} \times \frac{P}{m} + 2v = 50 - 17.660 = 32.340.$ And $V - \frac{V+1}{n} \times \overline{n-m-2v-1} \times \frac{P}{m} + 2v = 25 - \frac{26}{59} \times 32.340 = 10.748,$ the value of two joint lives whose ages are 27 and 38.

Note (N). Essay III. Page 237.

IT is plain that the purchaser of A's right, as stated in the first of the questions, to which this note refers, cannot get into possession, till the year when A and B shall be both dead; nor then, unless A happens to die *last*. Now, supposing the common complement of life n; the probability that A and B shall be *both* dead at the end of the *first* year, and A die last, is $\overline{1 - \frac{n-1}{n}} \times \overline{1 - \frac{n-1}{n}} \times \frac{1}{2} = \frac{1}{2} - \frac{n-1}{2n} - \frac{n-1}{2n} + \frac{\overline{n-1}^2}{2n^2}$.——In like manner, the probability that they shall be *both* dead at the end of the 2d, 3d, &c. years, and A survive, is $\frac{1}{2} - \frac{n-2}{2n} - \frac{n-2}{2n} + \frac{\overline{n-2}^2}{2n^2}$; $\frac{1}{2} - \frac{n-3}{2n} - \frac{n-3}{2n} + \frac{\overline{n-3}^2}{2n^2}$, &c. The *present* value, therefore, of the 1st, 2d, 3d, &c. rents of the reversionary estate is $\frac{1}{2r} - \frac{n-1}{2nr} - \frac{n-1}{2nr} + \frac{\overline{n-1}^2}{2nr}$, $\frac{1}{2r^2} - \frac{n-2}{2nr^2} - \frac{n-2}{2nr^2} + \frac{\overline{n-2}^2}{2n^2r^2}$, $\frac{1}{2r^3} - \frac{n-3}{2nr^3} - \frac{n-3}{2nr^3} + \frac{\overline{n-3}^2}{2n^2r^3}$, &c. Supposing r to signify 1*l*. increased by its interest for a year; and the estate to be 1*l*. *per annum*. And the *sum* of these terms continued *in infinitum*, is the value *required*.——But $\frac{1}{2r} + \frac{1}{2r^2} + \frac{1}{2r^3}$, &c. is *half* the

APPENDIX.

the perpetuity. And $\frac{n-1}{2nr} + \frac{n-1}{2nr} - \frac{\overline{n-1}|^2}{2n^2r} + \frac{n-2}{2nr^2} + \frac{n-2}{2nr^2} - \frac{\overline{n-2}|^2}{2n^2r^2} + \frac{n-3}{2nr^3} + \frac{n-3}{2nr^3} - \frac{\overline{n-3}|^2}{2n^2r^3}$, &c. is half the value of the *joint* lives, subtracted from *half* the sum of the values of the two *single* lives; that is, *half* the value of the *longest* of the two lives.

A similar demonstration may be applied to the other question.

Note (O). Essay IV. Page 267.

LET r be $1 l.$ increased by its interest for one year.

Let S represent any given interval of time, or number of years, during which the decrements of life in a table of observations continue equal.

a the number of the living in the table at the beginning of the first year of that interval.

b the number of the living in the table at the beginning of the year immediately following the same interval.

P the value of an annuity certain for S years.

p the value, in Table I. of $1 l.$ due at the end of S years.

Q the value, in Table VI. of an annuity for the life of a person whose age wants S years of 86.

N the value, in strict agreement with the given table of observations, of an annuity on the life of a person whose age is S years greater than the age at which the interval of equal decrements begins. Then,

$$Q + \frac{b}{a} \times \overline{P-Q}$$ will be the value, according to the table of observations, of an annuity for S years, on a life of the same age with that at which the interval of equal decrements begins. And

$$Q + \frac{b}{a} \times \overline{P-Q+pN}$$ will the value of an annuity on the whole duration of that life.

When S represents *one year*, Q vanishes, and the last expression becomes $\frac{b}{ar} \times \overline{1+N}$; which is the

rule

APPENDIX. 313

rule for finding, from the value given of any life, the value of a life one year younger.

These Theorems save much labour in calculating the values of life-annuities from tables of observations.

The first of them, with its investigation, may be found in page 341, 3d edition, of Mr. *De Moivre's Treatise on the Doctrine of Chances.* But it is necessary to observe, that the direction Mr. *De Moivre* has given for finding the value of Q is wrong. In consequence of calculating agreeably to this direction, he gives the value of a life at the age of 42, by Dr. *Halley's* table, greater than the value of the same life by his own hypothesis; whereas, it is evident, that the probabilities of living after 42, being all along less in Dr. *Halley's* table, than in the hypothesis, the *value* of the life must be also less.

The mathematical reader may easily satisfy himself, that the value of Q ought to be taken from Table VI. as I have directed.

An easy and accurate method of finding the values of single lives, agreeably to any given table of observations, is given by Mr. *Dodson* in his *Mathematical Repository,* vol. II. page 161.

There is also in Mr. *Simpson's Select Exercises,* page 275, a very easy rule for approximating to the values of single lives, according to Dr. *Halley's* table. But this rule must not be depended on; for I have found it half a year's purchase, and sometimes three-quarters of a year's purchase wrong.

To prevent the danger of mistaking the Theorem I have given, I have thought proper to subjoin the following example.

Let the table of observations be the *Breslaw* Table, or Table III. The value of a life at 78, by this

Table,

Table, is $\frac{49}{58r} + \frac{41}{58r^2} + \frac{34}{58r^3}$, &c. to the end of life. The number of terms in this series being small, it may be easily found to be 3.514, supposing interest at 4 *per cent.* and $\frac{1}{r}, \frac{1}{r^2}, \frac{1}{r^3}$, &c. being the values, in Table I. of 1 *l.* at the end of 1, 2, 3, &c. years.——From 78 to 74 the decrements of life continue equal; and therefore S=4. a=98. b=58. P= 3.6298, by Table II; p= .8548, by Table 1; Q= 1.406, by Table VI; N= 3.514. P—Q+pN = 5.227; and $\overline{Q + \frac{b}{a}}$ × $\overline{P—Q+pN}$ = 4.500, or the value of a life at 74.

From 74 to 70 there is another interval of equal decrements; and, by a like easy operation, the value of a life at 70 will be found to be 5.595.

TABLE

APPENDIX.

TABLE I.

The prefent Value of 1 *l*. to be received at the end of any number of years, not exceeding 100; difcounting at the rates of 3, 3½, 4, 4½, 5 and 6 *per cent.* compound intereft. *Computed yearly*

	3 per Ct.	3½ per Ct.	4 per Ct.	4½ per Ct.	5 per Ct.	6 per Ct.
1	,970874	,966184	,961538	,956938	,952381	,943396
2	,942596	,933511	,924556	,915730	,907029	,889996
3	,915142	,901943	,888996	,876297	,863838	,839619
4	,888487	,871442	,854804	,838561	,822702	,792094
5	,862609	,841973	,821927	,802451	,783526	,747258
6	,837484	,813501	,790315	,767896	,746215	,704961
7	,813092	,785991	,759918	,734828	,710681	,665057
8	,789409	,759412	,730690	,703185	,676839	,627412
9	,766417	,733731	,702587	,672904	,644609	,591898
10	,744094	,708919	,675564	,643928	,613913	,558395
11	,722421	,684946	,649581	,616199	,584679	,526788
12	,701380	,661783	,624597	,589664	,556837	,496969
13	,680951	,639404	,600574	,564272	,530321	,468839
14	,661118	,617782	,577475	,539973	,505068	,442301
15	,641862	,596891	,555265	,516720	,481017	,417265
16	,623167	,576706	,533908	,494469	,458112	,393646
17	,605016	,557204	,513373	,473176	,436297	,371364
18	,587395	,538361	,493628	,452800	,415521	,350344
19	,570286	,520156	,474642	,433302	,395734	,330513
20	,553676	,502566	,456387	,414643	,376889	,311805
21	,537549	,485571	,438834	,396787	,358942	,294155
22	,521893	,469151	,421955	,379701	,341850	,277505
23	,506692	,453286	,405726	,363350	,325571	,261797
24	,491934	,437957	,390121	,347703	,310068	,246979
25	,477606	,423147	,375117	,332731	,295303	,232999
26	,463695	,408838	,360689	,318402	,281241	,219810
27	,450189	,395012	,346817	,304691	,267848	,207368
28	,437077	,381654	,333477	,291571	,255094	,195630
29	,424346	,368748	,320651	,279015	,242946	,184557
30	,411987	,356278	,308319	,267000	,231377	,174110
31	,399987	,344230	,296460	,255502	,220359	,164255
32	,388337	,332590	,285053	,244500	,209866	,154957

APPENDIX

TABLE I. Continued.

	3 per Ct.	3½ per Ct.	4 per Ct.	4½ per Ct.	5 per Ct.	6 per Ct.
33	,377026	,321343	,274094	,233971	,199873	,146186
34	,366045	,310476	,263552	,223896	,190355	,137912
35	,355383	,299977	,253415	,214254	,181290	,130105
36	,345032	,289833	,243669	,205028	,172657	,122741
37	,334983	,280032	,234297	,196199	,164436	,115793
38	,325226	,270562	,225285	,187750	,156605	,109259
39	,315754	,261413	,216621	,179665	,149148	,103056
40	,306557	,252572	,208289	,171929	,142046	,097222
41	,297628	,244031	,200278	,164525	,135282	,091719
42	,288959	,235779	,192575	,157440	,128840	,086527
43	,280543	,227806	,185168	,150663	,122704	,081630
44	,272372	,220102	,178046	,144173	,116864	,077009
45	,264439	,212659	,171198	,137964	,111297	,072650
46	,256737	,205468	,164614	,132023	,105997	,068538
47	,249259	,198520	,158283	,126338	,100949	,064658
48	,241999	,191806	,152195	,120898	,096142	,060998
49	,234950	,185320	,146341	,115692	,091564	,057546
50	,228107	,179053	,140713	,110710	,087204	,054228
51	,221463	,172998	,135301	,105942	,083051	,051215
52	,215013	,167148	,130097	,101380	,079096	,048316
53	,208750	,161496	,125093	,097014	,075330	,045582
54	,202670	,156035	,120282	,092837	,071743	,043001
55	,196767	,150758	,115656	,088839	,068326	,040567
56	,191036	,145660	,111207	,085013	,065073	,038271
57	,185472	,140734	,106930	,081353	,061974	,036105
58	,180070	,135975	,102817	,077849	,059023	,034061
59	,174825	,131377	,098963	,074497	,056212	,032133
60	,169733	,126934	,095060	,071289	,053536	,030310
61	,164789	,122642	,091404	,068219	,050986	,028598
62	,159990	,118495	,087889	,065281	,048558	,026989
63	,155330	,114487	,084508	,062470	,046246	,025453
64	,150806	,110616	,081258	,059780	,044044	,024012
65	,146413	,106875	,078133	,057206	,041946	,022653
66	,142149	,103261	,075128	,054742	,039949	,021370
67	,138009	,099769	,072238	,052385	,038047	,020161
68	,133989	,096395	,069460	,050129	,036235	,019020
69	,130086	,093136	,066788	,047971	,034509	,017943
70	,126297	,019986	,064219	,045905	,032866	,016927

APPENDIX.

TABLE I. Continued.

	3 per Ct.	3½ per Ct.	4 per Ct.	4½ per Ct.	5 per Ct.	6 per Ct.
71	,122619	,086943	,061749	,043928	,031301	,015969
72	,119047	,084003	,059374	,042037	,029811	,015065
73	,115580	,081162	,057091	,040226	,028391	,014212
74	,112214	,078418	,054895	,038494	,027039	,013408
75	,108945	,075766	,052784	,036836	,025752	,012649
76	,105772	,073204	,050754	,035250	,024525	,011933
77	,102691	,070728	,048801	,033732	,023357	,011258
78	,099700	,068336	,046924	,032280	,022245	,010620
79	,096796	,066026	,045120	,030890	,021186	,010019
80	,093977	,063793	,043384	,029559	,020177	,009452
81	,091240	,061636	,041716	,028287	,019216	,008917
82	,088582	,059551	,040111	,027068	,018301	,008412
83	,086002	,057538	,038569	,025903	,017430	,007936
84	,083497	,055592	,037085	,024787	,016600	,007487
85	,081065	,053712	,035659	,023720	,015809	,007063
86	,078704	,051896	,034287	,022699	,015056	,006663
87	,076412	,050141	,032968	,021721	,014339	,006286
88	,074186	,048445	,031700	,020786	,013657	,005930
89	,072027	,046807	,030481	,019891	,013006	,005595
90	,069928	,045224	,029309	,019034	,012387	,005278
91	,067891	,043695	,028182	,018215	,011797	,004979
92	,065914	,042217	,027098	,017430	,011235	,004697
93	,063994	,040789	,026055	,016680	,010700	,004432
94	,062130	,039410	,025053	,015961	,010191	,004181
95	,060320	,038077	,024090	,015274	,009705	,003944
96	,058563	,036790	,023163	,014616	,009243	,003721
97	,056858	,035546	,022272	,013987	,008803	,003510
98	,055202	,034344	,021416	,013385	,008384	,003312
99	,053594	,033182	,020592	,012808	,007985	,003124
100	,052033	,032060	,019800	,012257	,007604	,002057

TABLE

TABLE II.

The present Value of an Annuity of One Pound, for any Number of Years not exceeding 100, at the several Rates of 3, 3½, 4, 5, and 6 *l. per Cent.*

Ye.	3 per Ct.	3½ per Ct.	4 per Ct.	5 per Ct.	6 per Ct.
1	.9708	.9662	.9615	.9523	.9433
2	1.9133	1.8997	1.8860	1.8594	1.8333
3	2.8286	2.8016	2.7750	2.7232	2.6730
4	3.7170	3.6731	3.6298	3.5459	3.4651
5	4.5797	4.5151	4.4518	4.3294	4.2123
6	5.4971	5.3286	5.2421	5.0756	4.9173
7	6.2302	6.1145	6.0020	5.7863	5.5823
8	7.0196	6.8740	6.7327	6.4632	6.2097
9	7.7861	7.6077	7.4353	7.1078	6.8016
10	8.5302	8.3166	8.1108	7.7212	7.3600
11	9.2526	9.0015	8.7604	8.3064	7.8868
12	9.9540	9.6633	9.3850	8.8632	8.3838
13	10.6349	10.3027	9.9856	9.3935	8.8526
14	11.2960	10.9205	10.5631	9.8986	9.2949
15	11.9379	11.5174	11.1183	10.3796	9.7122
16	12.5611	12.0941	11.6522	10.8377	10.1058
17	13.1661	12.6513	12.1656	11.2740	10.4772
18	13.7535	13.1897	12.6592	11.6895	10.8276
19	14.3238	13.7098	13.1339	12.0853	11.1581
20	14.8774	14.2124	13.5903	12.4622	11.4699
21	15.4150	14.6980	14.0291	12.8211	11.7640
22	15.9389	15.1671	14.4511	13.1630	12.0415
23	16.4436	15.6204	14.8568	13.4885	12.3033
24	16.9355	16.0584	15.2469	13.7986	12.5503
25	17.4131	16.4815	15.6220	14.0939	12.7833

TABLE II. Continued.

Ye.	3 per Ct.	3½ per Ct.	4 per Ct.	5 per Ct.	6 per Ct.
26	17.8768	16,8904	15.9827	14.3751	13.0031
27	18.3270	17.2854	16.3295	14.6430	13.2105
28	18.7641	17.6670	16.6630	14.8981	13.4061
29	19.1884	18.0358	16.9837	15.1410	13.5907
30	19.6004	18.3920	17.2920	15.3724	13.7648
31	20.0004	18.7363	17.5884	15.5928	13.9290
32	20.3887	19.0689	17.8735	15.8026	14.0840
33	20.7657	19.3902	18.1476	16.0025	14.2302
34	21.1318	19.7007	18.4111	16.1929	14.3681
35	21.4872	20.0007	18.6646	16.3741	14.4982
36	21.8322	20.2905	18.9082	16.5468	14.6209
37	22.1672	20.5705	19.1425	16.7112	14.7367
38	22.4924	20.8411	19.3678	16.8678	14.8460
39	22.8082	21.1025	19.5844	17.0170	14.9490
40	23.1147	21.3551	19.7927	17.1590	15.0462
41	23.4124	21.5991	19.9930	17.2943	15.1380
42	23.7013	21.8349	20.1856	17.4232	15.2245
43	23.9819	22.0627	20.3707	17.5459	15.3061
44	24.2542	22.2828	20.5488	17.6627	15.3831
45	24.5187	22.4955	20.7200	17.7740	15.4558
46	24.7754	22.7009	20.8846	17.8800	15.5243
47	25.0247	22.8994	21.0429	17.9810	15.5890
48	25.2667	23.0912	21.1951	18.0771	15.6500
49	25.5016	23.2766	21.3414	18.1687	15.7075
50	25.7297	23.4556	21.4821	18.2559	15.7618
51	25.9512	23.6286	21.6174	18.3389	15.8130
52	26.1662	23.7958	21.7475	18.4180	15.8613
53	26.3749	23.9573	21.8726	18.4934	15.9069
54	26.5776	24.1133	21.9929	18.5651	15.9499
55	26.7744	24.2641	22.1086	18.6334	15.9905

TABLE II. Continued.

Ye.	3 per Ct.	3½ per Ct.	4 per Ct.	5 per Ct.	6 per Ct.
56	26.9654	24.4097	22.2198	18.6985	16.0288
57	27.1509	24.5504	22.3267	18.7605	16.0649
58	27.3310	24.6864	22.4295	18.8195	16.0989
59	27.5058	24.8178	22.5284	18.8757	16.0311
60	27.6755	24.9447	22.6234	18.9292	16.1614
61	27.8403	25.0674	22.7148	18.9802	16.1900
62	28.0003	25.1859	22.8027	19.0288	16.2170
63	28.1556	25.3004	22.8872	19.0750	16.2424
64	28.3064	25.4110	22.9685	19.1191	16.2664
65	28.4528	25.5178	23.0466	19.1610	16.2891
66	28.5950	25.6211	23.1218	19.2010	16.3104
67	28.7330	25.7209	23.1940	19.2390	16.3306
68	28.8670	25.8173	23.2635	19.2753	16.3496
69	28.9971	25.9104	23.3302	19.3098	16.3676
70	29.1234	26.0004	23.3945	19.3426	16.3845
71	29.2460	26.0873	23.4562	19.3739	16.4005
72	29.3650	26.1713	23.5156	19.4037	16.4155
73	29.4806	26.2525	23.5727	19.4321	16.4297
74	29.5928	26.3309	23.6276	19.4592	16.4431
75	29.7018	26.4067	23.6804	19.4849	16.4558
76	29.8076	26.4799	23.7311	19.5094	16.4677
77	29.9102	26.5506	23.7799	19.5328	16.4790
78	30.0099	26.6190	23.8268	19.5550	16.4896
79	30.1067	26.6850	23.8720	19.5762	16.4996
80	30.2007	26.7488	23.9153	19.5964	16.5091
81	30.2920	26.8104	23.9571	19.6156	16.5180
82	30.3805	26.8700	23.9972	19.6339	16.5264
83	30.4665	26.9275	24.0357	19.6514	16.5343
84	30.5500	26.9831	24.0728	19.6680	16.5418
85	30.6311	27.0368	24.1085	19.6838	16.5489

TABLE II. Continued.

Ye.	3 per Ct.	3½ per Ct.	4 per Ct.	5 per Ct.	6 per Ct.
86	30.7098	27.0887	24.1428	19.6988	16.5556
87	30.7862	27.1388	24.1757	19.7132	16.5618
88	30.8604	27.1873	24.2074	19.7268	16.5678
89	30.9324	27.2341	24.2379	19.7398	16.5734
90	31.0024	27.2793	24.2672	19.7522	16.5786
91	31.0703	27.3230	24.2954	19.7640	16.5836
92	31.1362	27.3652	24.3225	19.7752	16.5883
93	31.2002	27.4060	24.3486	19.7859	16.5928
94	31.2623	27.4454	24.3736	19.7961	16.5969
95	31.3226	27.4835	24.3977	19.8058	16.6009
96	31.3812	27.5203	24.4209	19.8151	16.6046
97	31.4380	27.5558	24.4431	19.8239	16.6081
98	31.4932	27.5902	24.4646	19.8323	16.6114
99	31.5468	27.6234	24.4852	19.8403	16.6145
100	31.5989	27.6554	24.5050	19.8479	16.6175
Perpetuity.	33.3333	28.5714	25.0000	20.0000	16.6666

TABLE III.

Shewing the Probabilities of the Duration of Life, as deduced by Dr. *Halley* from Observations on the Bills of Mortality of BRESLAW.

Ages	Persons living.	Decr. of Life.	Ages.	Persons living.	Decr. of Life.	Ages.	Persons living.	Decr. of Life.
1	1000	145	31	523	8	61	232	10
2	855	57	32	515	8	62	222	10
3	798	38	33	507	8	63	212	10
4	760	28	34	499	9	64	202	10
5	732	22	35	490	9	65	192	10
6	710	18	36	481	9	66	182	10
7	692	12	37	472	9	67	172	10
8	680	10	38	463	9	68	162	10
9	670	9	39	454	9	69	152	10
10	661	8	40	445	9	70	142	11
11	653	7	41	436	9	71	131	11
12	646	6	42	427	10	72	120	11
13	640	6	43	417	10	73	109	11
14	634	6	44	407	10	74	98	10
15	628	6	45	397	10	75	88	10
16	622	6	46	387	10	76	78	10
17	616	6	47	377	10	77	68	10
18	610	6	48	367	10	78	58	9
19	604	6	49	357	11	79	49	8
20	598	6	50	346	11	80	41	7
21	592	6	51	335	11	81	34	6
22	586	7	52	324	11	82	28	5
23	579	6	53	313	11	83	23	4
24	573	6	54	302	10	84	19	4
25	567	7	55	292	10	85	15	4
26	560	7	56	282	10	86	11	3
27	553	7	57	272	10	87	8	3
28	546	7	58	262	10	88	5	2
29	539	8	59	252	10	89	3	2
30	531	8	60	242	10	90	1	1

APPENDIX.

TABLE IV.

Shewing the Probabilities of Life at NORTHAMPTON. See page 260, 261.

Ages.	Perſons living.	Decr. of Life.	Ages.	Perſons living.	Decr. of Life.	Ages.	Perſons living.	Decr. of Life.
0	1149	300	31	428	7	62	187	8
1	849	127	32	421	7	63	179	8
2	722	50	33	414	7	64	171	8
3	672	26	34	407	7	65	163	8
4	646	21	35	400	7	66	155	8
5	625	16	36	393	7	67	147	8
6	609	13	37	386	7	68	139	8
7	596	10	38	379	7	69	131	8
8	586	9	39	372	7	70	123	8
9	577	7	40	365	8	71	115	8
10	570	6	41	357	8	72	107	8
11	564	6	42	349	8	73	99	8
12	558	5	43	341	8	74	91	8
13	553	5	44	333	8	75	83	8
14	548	5	45	325	8	76	75	8
15	543	5	46	317	8	77	67	7
16	538	5	47	309	8	78	60	7
17	533	5	48	301	8	79	53	7
18	528	6	49	293	9	80	46	7
19	522	7	50	284	9	81	39	7
20	515	8	51	275	8	82	32	6
21	507	8	52	267	8	83	26	5
22	499	8	53	259	8	84	21	4
23	491	8	54	251	8	85	17	4
24	483	8	55	243	8	86	13	3
25	475	8	56	235	8	87	10	2
26	467	8	57	227	8	88	8	2
27	459	8	58	219	8	89	6	2
28	451	8	59	211	8	90	4	2
29	443	8	60	203	8	91	2	1
30	435	7	61	195	8	92	1	1

APPENDIX.
TABLE V.

Shewing the Probabilities of Life at NORWICH.
See page 262.

Ages.	Persons living.	Decr. of Life.	Ages.	Persons living.	Decr. of Life.	Ages.	Persons living.	Decr. of Life.
0	1185	320	32	392	6	63	174	9
1	865	160	33	386	6	64	165	9
2	705	60	34	380	6	65	156	9
3	645	32	35	374	6	66	147	9
4	613	23	36	368	6	67	138	9
5	590	20	37	362	6	68	129	9
6	570	16	38	356	6	69	120	9
7	554	13	39	350	7	70	111	9
8	541	11	40	343	6	71	102	8
9	530	9	41	337	6	72	94	8
10	521	7	42	331	6	73	86	8
11	514	6	43	325	7	74	78	8
12	508	6	44	318	7	75	70	8
13	502	5	45	311	7	76	62	7
14	497	5	46	304	7	77	55	7
15	492	5	47	297	7	78	48	6
16	487	5	48	290	7	79	42	5
17	482	5	49	283	7	80	37	5
18	477	5	50	276	7	81	32	4
19	472	5	51	269	7	82	28	4
20	467	6	52	262	7	83	24	4
21	461	6	53	255	8	84	20	3
22	455	6	54	247	8	85	17	3
23	449	6	55	239	8	86	14	3
24	443	6	56	231	8	87	11	2
25	437	6	57	223	8	88	9	2
26	431	7	58	215	8	89	7	2
27	424	7	59	207	8	90	5	2
28	417	7	60	199	8	91	3	2
29	410	6	61	191	8	92	1	1
30	404	6	62	183	9	93		
31	398	6						

APPENDIX.

TABLE VI. (a).

Shewing the present Values of an Annuity of 1 *l.* on a Single Life, according to Mr. *De Moivre's* hypothesis; and, therefore, nearly, according to the probabilities of life at Breslaw, Norwich, and Northampton. See p. 2, and p. 267.

Age.	3 per Ct.	3½ per Ct.	4 per Ct.	4½ per Ct.	5 per Ct.	6 per Ct.
8	19,736	18,160	16,791	15,595	14,544	12,790
9	19,868	18,269	16,882	15,672	14,607	12,839
10	19,868	18,269	16,882	15,672	14,607	12,839
11	19,736	18,160	16,791	15,595	14,544	12,790
12	19,604	18,049	16,698	15,517	14,480	12,741
13	19,469	17,937	16,604	15,437	14,412	12,691
14	19,331	17,823	16,508	15,356	14,342	12,639
15	19,192	17,707	16,410	15,273	14,271	12,586
16	19,050	17,588	16,311	15,189	14,197	12,532
17	18,905	17,467	16,209	15,102	14,123	12,476
18	18,759	17,344	16,105	15,015	14,047	12,419
19	18,610	17,220	15,999	14,923	13,970	12,361
20	18,458	17,093	15,891	14,831	13,891	12,301
21	18,305	16,963	15,781	14,737	13,810	12,239
22	18,148	16,830	15,669	14,641	13,727	12,177
23	17,990	16,696	15,554	14,543	13,642	12,112
24	17,827	16,559	15,437	14,442	13,555	12,045
25	17,664	16,419	15,318	14,340	13,466	11,978
26	17,497	16,277	15,197	14,235	13,375	11,908
27	17,327	16,133	15,073	14,128	13,282	11,837
28	17,154	15,985	14,946	14,018	13,186	11,763
29	16,979	15,835	14,816	13,905	13,088	11,688
30	16,800	15,682	14,684	13,791	12,988	11,610
31	16,620	15,526	14,549	13,673	12,855	11,530
32	16,436	15,367	14,411	13,553	12,780	11,449
33	16,248	15,204	14,270	13,430	12,673	11,365

(a) This Table is the same with Mr. *De Moivre's* Table of the values of single lives, published in his *Treatise on Life Annuities*, and carried as far as the age of 79, to three places of decimals, by Mr. *Dodson* in his *Mathematical Repository*, vol. II. p. 169.

TABLE VI. Continued.

Age.	3 per Ct.	3½ per Ct.	4 per Ct.	4½ per Ct.	5 per Ct.	6 per Ct.
34	16,057	15,039	14,126	13,304	12,562	11,278
35	15,864	14,871	13,979	13,175	12,449	11,189
36	15,666	14,699	13,829	13,044	12,333	11,098
37	15,465	14,524	13,676	12,909	12,214	11,003
38	15,260	14,345	13,519	12,771	12,091	10,907
39	15,053	14,163	13,359	12,630	11,966	10,807
40	14,842	13,978	13,196	12,485	11,837	10,704
41	14,626	13,789	13,028	12,337	11,705	10,599
42	14,407	13,596	12,858	12,185	11,570	10,490
43	14,185	13,399	12,683	12,029	11,431	10,378
44	13,958	13,199	12,504	11,870	11,288	10,263
45	13,728	12,993	12,322	11,707	11,142	10,144
46	13,493	12,784	12,135	11,540	10,992	10,021
47	13,254	12,571	11,944	11,368	10,837	9,895
48	13,012	12,354	11,748	11,192	10,679	9,765
49	12,764	12,131	11,548	11,012	10,515	9,630
50	12,511	11,904	11,344	10,827	10,348	9,492
51	12,255	11,673	11,135	10,638	10,176	9,349
52	11,994	11,437	10,921	10,443	9,999	9,201
53	11,729	11,195	10,702	10,243	9,817	9,049
54	11,457	10,950	10,478	10,039	9,630	8,891
55	11,183	10,698	10,248	9,829	9,437	8,729
56	10,902	10,443	10,014	9,614	9,239	8,561
57	10,616	10,181	9,773	9,393	9,036	8,387
58	10,325	9,913	9,527	9,166	8,826	8,208
59	10,029	9,640	9,275	8,933	8,611	8,023
60	9,727	9,361	9,017	8,694	8,389	7,831
61	9,419	9,076	8,753	8,449	8,161	7,633
62	9,107	8,786	8,482	8,197	7,926	7,428
63	8,787	8,488	8,205	7,938	7,684	7,216
64	8,462	8,185	7,921	7,672	7,435	6,997
65	8,132	7,875	7,631	7,399	7,179	6,770
66	7,794	7,558	7,333	7,119	6,915	6,535
67	7,450	7,234	7,027	6,831	6,643	6,292
68	7,099	6,902	6,714	6,534	6,362	6,040
69	6,743	6,565	6,394	6,230	6,073	5,779
70	6,378	6,219	6,065	5,918	5,775	5,508

TABLE VI. Continued.

Age.	3 per Ct.	3½ per Ct.	4 per Ct.	4½ per Ct.	5 per Ct.	6 per Ct.
71	6,008	5,865	5,728	5,596	5,468	5,228
72	5,631	5,505	5,383	5,265	5,152	4,937
73	5,246	5,136	5,029	4,926	4,826	4,636
74	4,854	4,759	4,666	4,576	4,489	4,324
75	4,453	4,373	4,293	4,217	4,143	4,000
76	4,046	3,978	3,912	3,847	3,784	3,664
77	3,632	3,575	3,520	3,467	3,415	3,315
78	3,207	3,163	3,111	3,076	3,034	2,953
79	2,776	2,741	2,707	2,673	2,641	2,578
80	2,334	2,309	2,284	2,259	2,235	2,188
81	1,886	1,867	1,850	1,832	1,816	1,783
82	1,429	1,411	1,406	1,394	1,384	1,362
83	0,961	0,955	0,950	0,943	0,937	0,925
84	0,484	0,483	0,481	0,479	0,476	0,472
85	0,000	0,000	0,000	0,000	0,000	0,000

APPENDIX.

TABLE VII.

Shewing the Value of an Annuity on the joint continuance of Two Lives, according to Mr. *De Moivre's Hypothesis*; and, therefore, nearly according to the probabilities of life at BRESLAW, NORWICH, and NORTHAMPTON. See Essay II. and p. 2, 3, 231, 267.

Age of the youngest.	Age of the eldest.	Value at 3 per Cent.	Value at 4 per Cent.	Value at 5 per Cent.
10	10	15.206	13.342	11.855
	15	14.878	13.093	11.661
	20	14.503	12.808	11.430
	25	14.074	12.480	11.182
	30	13.585	12.102	10.884
	35	13.025	11.665	10.537
	40	12.381	11.156	10.128
	45	11.644	10.564	9.646
	50	10.796	9.871	9.074
	55	9.822	9.059	8.391
	60	8.704	8.105	7.572
	65	7.417	6.980	6.585
	70	5.936	5.652	5.391
15	15	14.574	12.860	11.478
	20	14.225	12.593	11.266
	25	13.822	12.281	11.022
	30	13.359	11.921	10.736
	35	12.824	11.501	10.402
	40	12.207	11.013	10.008
	45	11.496	10.440	9.541
	50	10.675	9.767	8.985
	55	9.727	8.975	8.318
	60	8.632	8.041	7.515
	65	7.377	6.934	6.544
	70	5.932	5.623	5.364

APPENDIX.

TABLE VII. Continued.

Age of the youngest.	Age of the eldest.	Value at 3 per Cent.	Value at 4 per Cent.	Value at 5 per Cent.
	20	13.904	a 12.341	11.067
	25	13.531	12.051	10.840
	30	13.098	11.711	10.565
	35	12.594	11.314	10.278
	40	12.008	10.847	9.870
20	45	11.325	10.297	9.420
	50	10.536	9.648	8.880
	55	9.617	8.879	8.233
	60	8.549	7.967	7.448
	65	7.308	6.882	6.495
	70	5.868	5.590	5.333
	25	13.192	11.786	10.621
	30	12.794	11.468	10.367
	35	12.333	11.095	10.067
	40	11.776	10.655	9.708
25	45	11.130	10.131	9.278
	50	10.374	9.509	8.761
	55	9.488	8.766	8.134
	60	8.452	7.880	7.371
	65	7.241	6.826	6.440
	70	5.826	5.551	5.294
	30	12.434	11.182	10.133
	35	12.010	10.838	9.854
	40	11.502	10.428	9.514
	45	10.898	9.936	9.112
30	50	10.183	9.345	8.620
	55	9.338	8.634	8.018
	60	8.338	7.779	7.280
	65	7.161	6.748	6.373
	70	5.777	5.505	5.254

(a) by Norwich
is 13.066

APPENDIX.

TABLE VII. Continued.

Age of the youngest.	Age of the eldest.	Value at 3 per Cent.	Value at 4 per Cent.	Value at 5 per Cent.
35	35	11.632	10.530	9.600
	40	11.175	10.157	9.291
	45	10.622	9.702	8.913
	50	9.955	9.149	8.450
	55	9.156	8.476	7.879
	60	8.202	7.658	7.172
	65	7.066	6.662	6.294
	70	5.718	5.450	5.203
40	40	10.777	9.826	9.014
	45	10.283	9.418	8.671
	50	9.677	8.911	8.244
	55	8.936	8.283	7.710
	60	8.038	7.510	7.039
	65	6.951	6.556	6.198
	70	5.646	5.383	5.141
45	45	9.863	9.063	8.370
	50	9.331	8.619	7.987
	55	8.662	8.044	7.500
	60	7.831	7.332	6.875
	65	6.807	6.425	6.080
	70	5.556	5.300	5.063
50	50	8.892	8.235	7.660
	55	8.312	7.738	7.230
	60	7.568	7.091	6.664
	65	6.623	6.258	5.926
	70	5.442	5.193	4.964
55	55	7.849	7.332	6.873
	60	7.220	6.781	6.386
	65	6.379	6.036	5.724
	70	5.291	5.053	4.833

TABLE VII. Continued.

Age of the youngest.	Age of the eldest.	Value at 3 per Cent.	Value at 4 per Cent.	Value at 5 per Cent.
60	60	6.737	6.351	6.001
	65	6.043	5.730	5.444
	70	5.081	4.858	4.653
65	65	5.547	5.277	5.031
	70	4.773	4.571	4.385
70	70	4.270	4.104	3.952

TABLE

TABLE VIII.

Shewing the Probability of the Duration of Life in LONDON, deduced by Mr. *Simpson* from observations on the bills of mortality in LONDON for 10 years, from 1728 to 1737.

Ages.	Persons living.	Decr. of Life.	Ages.	Persons living.	Decr. of Life.	Ages.	Persons living.	Decr. of Life.
0	1000	320	27	321	6	54	135	6
1	680	133	28	315	7	55	129	6
2	547	51	29	308	7	56	123	6
3	496	27	30	301	7	57	117	5
4	469	17	31	294	7	58	112	5
5	452	12	32	287	7	59	107	5
6	440	10	33	280	7	60	102	5
7	430	8	34	273	7	61	97	5
8	422	7	35	266	7	62	92	5
9	415	5	36	259	7	63	87	5
10	410	5	37	252	7	64	82	5
11	405	5	38	245	8	65	77	5
12	400	5	39	237	8	66	72	5
13	395	5	40	229	7	67	67	5
14	390	5	41	222	8	68	62	4
15	385	5	42	214	8	69	58	4
16	380	5	43	206	7	70	54	4
17	375	5	44	199	7	71	50	4
18	370	5	45	192	7	72	46	4
19	365	5	46	185	7	73	42	3
20	360	5	47	178	7	74	39	3
21	355	5	48	171	6	75	36	3
22	350	5	49	165	6	76	33	3
23	345	6	50	159	6	77	30	3
24	339	6	51	153	6	78	27	2
25	333	6	52	147	6	79	25	
26	327	6	53	141	6			

APPENDIX.

TABLE IX.

Shewing the *Expectations* of Life in LONDON, according to the preceding Table. See Mr. *Simpson's Select Exercises*, p. 255.

Age.	Expectation.	Age.	Expectation.	Age.	Expectation.
1	27.0	28	24.6	55	14.2
2	32.0	29	24.1	56	13.8
3	34.0	30	23.6	57	13.4
4	35.6	31	23.1	58	13.1
5	36.0	32	22.7	59	12.7
6	36.0	33	22.3	60	12.4
7	35.8	34	21.9	61	12.0
8	35.6	35	21.5	62	11.6
9	35.2	36	21.1	63	11.2
10	34.8	37	20.7	64	10.8
11	34.3	38	20.3	65	10.5
12	33.7	39	19.9	66	10.1
13	33.1	40	19.6	67	9.8
14	32.5	41	19.2	68	9.4
15	31.9	42	18.8	69	9.1
16	31.3	43	18.5	70	8.8
17	30.7	44	18.1	71	8.4
18	30.1	45	17.8	72	8.1
19	29.5	46	17.4	73	7.8
20	28.9	47	17.0	74	7.5
21	28.3	48	16.7	75	7.2
22	27.7	49	16.3	76	6.8
23	27.2	50	16.0	77	6.4
24	26.6	51	15.6	78	6.0
25	26.1	52	15.2	79	5.5
26	25.6	53	14.9	80	5.0
27	25.1	54	14.5		

334 APPENDIX.

TABLE X.

Shewing the Value of an Annuity on *One* Life, according to the Probabilities of Life in LONDON, See Mr. *Simpson's Select Exercises*, p. 260.

Age.	Yrs. purchase at 3 per Cent.	Yrs. purchase at 4 per Cent.	Yrs. purchase at 5 per Cent.	Age.	Yrs. purchase at 3 per Cent.	Yrs. purchase at 4 per Cent.	Yrs. purchase at 5 per Cent.	Age.	Yrs. purchase at 3 per Cent.	Yrs. purchase at 4 per Cent.	Yrs. purchase at 5 per Cent.
6	18.8	16.2	14.1	31	14.8	12.9	11.4	56	10.1	9.1	8.4
7	18.9	16.3	14.2	32	14.6	12.7	11.3	57	9.9	8.9	8.2
8	19.0	16.4	14.3	33	14.4	12.6	11.2	58	9.6	8.7	8.1
9	19.0	16.4	14.3	34	14.2	12.4	11.0	59	9.4	8.6	8.0
10	19.0	16.4	14.3	35	14.1	12.3	10.9	60	9.2	8.4	7.9
11	19.0	16.4	14.3	36	13.9	12.1	10.8	61	8.9	8.2	7.7
12	18.9	16.3	14.2	37	13.7	11.9	10.6	62	8.7	8.1	7.6
13	18.7	16.2	14.1	38	13.5	11.8	10.5	63	8.5	7.9	7.4
14	18.5	16.0	14.0	39	13.3	11.6	10.4	64	8.3	7.7	7.3
15	18.3	15.8	13.9	40	13.2	11.5	10.3	65	8.0	7.5	7.1
16	18.1	15.6	13.7	41	13.0	11.4	10.2	66	7.8	7.3	6.9
17	17.9	15.4	13.5	42	12.8	11.2	10.1	67	7.6	7.1	6.7
18	17.6	15.2	13.4	43	12.6	11.1	10.0	68	7.4	6.9	6.6
19	17.4	15.0	13.2	44	12.5	11.0	9.9	69	7.1	6.7	6.4
20	17.2	14.8	13.0	45	12.3	10.8	9.8	70	6.9	6.5	6.2
21	17.0	14.7	12.9	46	12.1	10.7	9.7	71	6.7	6.3	6.0
22	16.8	14.5	12.7	47	11.9	10.5	9.5	72	6.5	6.1	5.8
23	16.5	14.3	12.6	48	11.8	10.4	9.4	73	6.2	5.9	5.6
24	16.3	14.1	12.4	49	11.6	10.2	9.3	74	5.9	5.6	5.4
25	16.1	14.0	12.3	50	11.4	10.1	9.2	75	5.6	5.4	5.2
26	15.9	13.8	12.1	51	11.2	9.9	9.0				
27	15.6	13.6	12.0	52	11.0	9.8	8.9				
28	15.4	13.4	11.8	53	10.7	9.6	8.8				
29	15.2	13.2	11.7	54	10.5	9.4	8.6				
30	15.0	13.1	11.6	55	10.3	9.3	8.5				

APPENDIX. 333

TABLE XI.

Shewing the Value of an Annuity on the joint continuance of Two Lives, according to the probabilities of Life in LONDON. See Mr. *Simpson's Select Exercises*, p. 266.

Age of the youngest.	Age of the eldest.	Value at 3 per Cent.	Value at 4 per Cent.	Value at 5 per Cent.	Age of the youngest.	Age of the eldest.	Value at 3 per Cent.	Value at 4 per Cent.	Value at 5 per Cent.
	10	14.7	13.0	11.6		20	12.8	11.3	10.1
	15	14.3	12.7	11.3		25	12.2	10.8	9.7
	20	13.8	12.2	10.8		30	11.6	10.3	9.2
	25	13.1	11.6	10.2		35	10.9	9.8	8.8
	30	12.3	10.9	9.7		40	10.2	9.2	8.4
10	35	11.5	10.2	9.1	20	45	9.5	8.6	7.9
	40	10.7	9.6	8.6		50	8.8	8.0	7.4
	45	10.0	9.0	8.1		55	8.1	7.5	6.9
	50	9.3	8.4	7.6		60	7.4	6.9	6.4
	55	8.6	7.8	7.1		65	6.7	6.3	5.9
	60	7.8	7.2	6.6		70	6.0	5.7	5.4
	65	6.9	6.5	6.1		75	5.2	5.0	4.8
	70	6.1	5.8	5.5					
	75	5.3	5.1	4.9		25	11.8	10.5	9.4
						30	11.3	10.1	9.0
	15	13.9	12.3	11.0		35	10.7	9.6	8.6
	20	13.3	11.8	10.5		40	10.0	9.1	8.2
	25	12.6	11.2	10.1		45	9.4	8.5	7.8
	30	11.9	10.6	9.5	25	50	8.7	7.9	7.3
	35	11.2	10.0	9.0		55	8.0	7.4	6.8
	40	10.4	9.4	8.5		60	7.3	6.8	6.3
15	45	9.6	8.8	8.0		65	6.6	6.2	5.8
	50	8.9	8.2	7.5		70	5.9	5.6	5.3
	55	8.2	7.6	7.0		75	5.1	4.9	4.7
	60	7.5	7.0	6.5					
	65	6.8	6.4	6.0		30	10.8	9.6	8.6
	70	6.0	5.7	5.4	30	35	10.3	9.2	8.3
	75	5.2	5.0	4.8		40	9.7	8.8	8.0

TABLE XI. Continued.

Age of the youngest.	Age of the eldest.	Value at 3 per Cent.	Value at 4 per Cent.	Value at 5 per Cent.	Age of the youngest.	Age of the eldest.	Value at 3 per Cent.	Value at 4 per Cent.	Value at 5 per Cent.
30	45	9.1	8.3	7.6	45	65	6.3	5.8	5.4
30	50	8.5	7.8	7.2	45	70	5.6	5.3	5.0
30	55	7.9	7.3	6.7	45	75	4.9	4.7	4.5
30	60	7.2	6.7	6.2					
30	65	6.5	6.1	5.7	50	50	7.6	6.8	6.2
30	70	5.8	5.5	5.2	50	55	7.2	6.5	6.0
30	75	5.1	4.9	4.7	50	60	6.7	6.1	5.7
					50	65	6.2	5.7	5.3
35	35	9.9	8.8	8.0	50	70	5.5	5.2	4.9
35	40	9.4	8.5	7.7	50	75	4.8	4.6	4.4
35	45	8.9	8.1	7.4					
35	50	8.3	7.6	7.0	55	55	6.9	6.2	5.7
35	55	7.7	7.1	6.6	55	60	6.5	5.9	5.5
35	60	7.1	6.5	6.1	55	65	6.0	5.6	5.2
35	65	6.4	6.0	5.6	55	70	5.4	5.1	4.8
35	70	5.7	5.4	5.1	55	75	4.7	4.5	4.3
35	75	5.0	4.8	4.6					
					60	60	6.1	5.6	5.2
40	40	9.1	8.1	7.3	60	65	5.7	5.3	4.9
40	45	8.7	7.8	7.1	60	70	5.2	4.9	4.6
40	50	8.2	7.4	6.8	60	75	4.6	4.4	4.2
40	55	7.6	6.9	6.4					
40	60	7.0	6.4	6.0	65	65	5.4	5.0	4.7
40	65	6.4	5.9	5.5	65	70	4.9	4.6	4.4
40	70	5.7	5.4	5.1	65	75	4.4	4.2	4.0
40	75	5.0	4.8	4.6					
					70	70	4.6	4.4	4.2
45	45	8.3	7.4	6.7	70	75	4.2	4.0	3.9
45	50	7.9	7.1	6.5					
45	55	7.4	6.7	6.2	75	75	3.8	3.7	3.6
45	60	6.8	6.3	5.8					

APPENDIX.
TABLE XII.

Shewing the Probabilities of Life in LONDON, on the fuppofition, that all who die in LONDON were born there. Formed from the Bills, for 10 years, from 1759 to 1768. See p. 250.

Ages.	Perfons living.	Decr. of Life.	Ages.	Perfons living.	Decr. of Life.	Ages.	Perfons living.	Decr. of Life.
0	1000	240	31	404	9	62	132	7
1	760	99	32	395	9	63	125	7
2	661	42	33	386	9	64	118	7
3	619	29	34	377	9	65	111	7
4	590	21	35	368	9	66	104	7
5	569	11	36	359	9	67	97	7
6	558	10	37	350	9	68	90	7
7	548	7	38	341	9	69	83	7
8	541	6	39	332	10	70	76	6
9	535	5	40	322	10	71	70	6
10	530	4	41	312	10	72	64	6
11	526	4	42	302	10	73	58	5
12	522	4	43	292	10	74	53	5
13	518	3	44	282	10	75	48	5
14	515	3	45	272	10	76	43	5
15	512	3	46	262	10	77	38	5
16	509	3	47	252	10	78	33	4
17	506	3	48	242	9	79	29	4
18	503	4	49	233	9	80	25	3
19	499	5	50	224	9	81	22	3
20	494	7	51	215	9	82	19	3
21	487	8	52	206	8	83	16	3
22	479	8	53	198	8	84	13	2
23	471	8	54	190	7	85	11	2
24	463	8	55	183	7	86	9	2
25	455	8	56	176	7	87	7	2
26	447	8	57	169	7	88	5	1
27	439	8	58	162	7	89	4	1
28	431	9	59	155	8	90	3	1
29	422	9	60	147	8			
30	413	9	61	139	7			

TABLE XIII.

Shewing the *true* Probabilities of Life in LONDON 'till the Age of 19. See p. 254.

Age.	Persons living.	Decrements of Life.
0	750	240
1	510	99
2	411	42
3	369	29
4	340	21
5	319	11
6	308	10
7	298	7
8	291	6
9	285	5
10	280	4
11	276	4
12	272	4
13	268	3
14	265	3
15	262	3
16	259	3
17	256	3
18	253	4
19	249	
20	494	
21	487	
&c.	&c.	

The numbers in the second column to be continued as in the last Table.

APPENDIX.
TABLE XIV.

Shewing the *true* Probabilities of Life in LONDON for all Ages. Formed from the Bills for 10 years, from 1759 to 1768. See p. 256.

Ages.	Persons living.	Decr. of Life.	Ages.	Persons living.	Decr. of Life.	Ages.	Persons living.	Decr. of Life.
0	1518	486	31	404	9	62	132	7
1	1032	200	32	395	9	63	125	7
2	832	85	33	386	9	64	118	7
3	747	59	34	377	9	65	111	7
4	688	42	35	368	9	66	104	7
5	646	23	36	359	9	67	97	7
6	623	20	37	350	9	68	90	7
7	603	14	38	341	9	69	83	7
8	589	12	39	332	10	70	76	6
9	577	10	40	322	10	71	70	6
10	567	9	41	312	10	72	64	6
11	558	9	42	302	10	73	58	5
12	549	8	43	292	10	74	53	5
13	541	7	44	282	10	75	48	5
14	534	6	45	272	10	76	43	5
15	528	6	46	262	10	77	38	5
16	522	7	47	252	10	78	33	4
17	515	7	48	242	9	79	29	4
18	508	7	49	233	9	80	25	3
19	501	7	50	224	9	81	22	3
20	494	7	51	215	9	82	19	3
21	487	8	52	206	8	83	16	3
22	479	8	53	198	8	84	13	2
23	471	8	54	190	7	85	11	2
24	463	8	55	183	7	86	9	2
25	455	8	56	176	7	87	7	2
26	447	8	57	169	7	88	5	1
27	439	8	58	162	7	89	4	1
28	431	9	59	155	8	90	3	1
29	422	9	60	147	8			
30	413	9	61	139	7			

TABLE XV.

Shewing the Value of an Annuity on the longest of Two given Lives, according to the Probabilities of Life in LONDON. See Mr. *Simpson's Select Exercises*, p. 268.

Age of the youngest	Age of the eldest	Value at 5 per Cent.	Value at 4 per Cent.	Value at 3 per Cent.	Age of the youngest	Age of the eldest	Value at 5 per Cent.	Value at 4 per Cent.	Value at 3 per Cent.
10	10	17.1	19.9	23.4		20	15.8	18.3	21.6
	15	16.8	19.5	22.9		25	15.5	17.9	21.1
	20	16.6	19.1	22.5		30	15.3	17.6	20.7
	25	16.4	18.8	22.2		35	15.1	17.4	20.4
	30	16.2	18.6	21.9		40	15.0	17.2	20.1
	35	16.1	18.4	21.6		45	14.9	17.0	19.9
	40	16.0	18.3	21.4	20	50	14.7	16.8	19.6
	45	15.9	18.2	21.2		55	14.5	16.6	19.4
	50	15.8	18.0	20.9		60	14.3	16.3	19.1
	55	15.7	17.8	20.7		65	14.1	16.0	18.7
	60	15.5	17.6	20.4		70	13.8	15.7	18.2
	65	15.3	17.4	20.1		75	13.5	15.3	17.7
	70	15.1	17.2	19.8		25	15.1	17.4	20.3
	75	14.8	16.9	19.5		30	14.9	17.0	19.8
	15	16.7	19.3	22.8		35	14.7	16.7	19.4
	20	16.4	18.9	22.3		40	14.5	16.5	19.2
	25	16.2	18.6	21.9		45	14.3	16.3	18.9
	30	16.0	18.3	21.6	25	50	14.2	16.1	18.7
	35	15.9	18.1	21.3		55	14.0	15.9	18.4
	40	15.7	17.9	21.1		60	13.8	15.6	18.0
15	45	15.6	17.8	20.9		65	13.6	15.3	17.6
	50	15.4	17.6	20.7		70	13.3	15.0	17.2
	55	15.3	17.4	20.4		75	12.9	14.6	16.7
	60	15.2	17.2	20.1		30	14.5	16.6	19.3
	65	15.0	16.9	19.8	30	35	14.2	16.2	18.8
	70	14.7	16.6	19.4		40	14.0	15.9	18.4
	75	14.4	16.3	18.9		45	13.8	15.6	18.1

APPENDIX. 341

TABLE XV. Continued.

Age of the youngest.	Age of the eldest.	Value at 5 per Cent.	Value at 4 per Cent.	Value at 3 per Cent.	Age of the youngest.	Age of the eldest.	Value at 5 per Cent.	Value at 4 per Cent.	Value at 3 per Cent.
30	50	13.6	15.4	17.8		65	11.4	12.5	14.1
	55	13.4	15.1	17.4	45	70	11.0	12.0	13.6
	60	13.2	14.8	17.0		75	10.6	11.6	13.1
	65	12.9	14.5	16.6					
	70	12.6	14.1	16.1		50	12.1	13.3	15.0
	75	12.2	13.7	15.6		55	11.7	12.9	14.5
						60	11.3	12.4	13.9
35	35	13.8	15.8	18.3	50				
	40	13.5	15.4	17.8		65	10.9	12.0	13.3
	45	13.3	15.1	17.4		70	10.5	11.5	12.8
	50	13.1	14.8	17.1		75	10.1	11.0	12.3
	55	12.9	14.5	16.7					
	60	12.7	14.2	16.3		55	11.3	12.4	13.6
	65	12.4	13.8	15.8	55	60	10.9	11.9	13.0
	70	12.0	13.4	15.3		65	10.5	11.3	12.4
	75	11.6	13.0	14.8		70	10.0	10.8	11.8
						75	9.5	10.3	11.3
40	40	13.3	15.0	17.3		60	10.5	11.2	12.2
	45	13.0	14.6	16.8	60	65	10.0	10.6	11.5
	50	12.7	14.2	16.3		70	9.5	10.1	10.9
	55	12.4	13.9	15.9		75	9.0	9.5	10.3
	60	12.1	13.5	15.4		65	9.4	10.0	10.7
	65	11.8	13.1	14.9	65	70	8.9	9.4	10.0
	70	11.4	12.7	14.5		75	8.3	8.7	9.3
	75	11.0	12.3	14.0					
					70	70	8.2	8.6	9.2
45	45	12.8	14.2	16.2		75	7.6	7.9	8.4
	50	12.5	13.8	15.7					
	55	12.1	13.4	15.2	75	75	6.9	7.2	7.6
	60	11.7	12.9	14.7					

TABLE XVI.

Shewing the Value of an Annuity on the longest of Two Lives, according to Mr. *De Moivre's Hypothesis*; and, therefore, nearly according to the probabilities of Life at BRESLAW, NORWICH, and NORTHAMPTON. See Page 231, 267, 268.

Age of the youngest	Age of the eldest	Value at 3 per Cent.	Value at 4 per Cent.	Value at 5 per Cent.	Age of the youngest	Age of the eldest	Value at 3 per Cent.	Value at 4 per Cent.	Value at 5 per Cent.
10	10	24.53	20.42	17.36		20	23.01	19.44	16.79
	15	24.18	20.20	17.22		25	22.59	19.16	16.52
	20	23.82	19.96	17.07		30	22.16	18.86	16.31
	25	23.45	19.72	16.89		35	21.73	18.55	16.06
	30	23.08	19.46	16.71	20	40	21.29	18.24	15.86
	35	22.71	19.20	16.52		45	20.86	17.92	15.61
	40	22.33	18.92	16.31		50	20.43	17.59	15.36
	45	21.95	18.64	16.10		55	20.02	17.26	15.10
	50	21.58	18.35	15.88		60	19.63	16.94	14.83
	55	21.23	18.07	15.65		65	19.28	16.64	14.57
	60	20.89	17.79	15.42		70	18.97	16.31	14.33
	65	20.58	17.53	15.20		25	22.14	19.85	16.31
	70	20.31	17.30	14.99		30	21.67	18.53	16.09
	15	23.81	19.96	17.06		35	21.20	18.20	15.85
	20	23.42	19.71	16.89	25	40	20.73	17.86	15.59
	25	23.03	19.45	16.71		45	20.26	17.51	15.33
	30	22.63	19.17	16.52		50	19.80	17.15	15.05
	35	22.23	18.89	16.32		55	19.36	16.80	14.77
15	40	21.83	18.59	16.10		60	18.94	16.45	14.48
	45	21.42	18.29	15.87		65	18.55	16.12	14.20
	50	21.03	18.00	15.63		70	18.22	15.83	13.95
	55	20.65	17.68	15.40					
	60	20.29	17.38	15.15					
	65	19.95	17.11	14.91					
	70	19.64	16.85	14.68					

APPENDIX.

TABLE XVI. Continued.

Age of the eldeſt.	Value at 3 per Cent.	Value at 4 per Cent.	Value at 5 per Cent.	Age of the youngeſt.	Age of the eldeſt.	Value at 3 per Cent.	Value at 4 per Cent.	Value at 5 per Cent.
30	21.16	18.18	15.84					
35	20.65	17.82	15.58		50	16.13	14.45	13.03
40	20.14	17.45	15.31		55	15.38	13.85	12.55
45	19.63	17.07	15.02	50	60	14.67	13.27	12.07
50	19.13	16.68	14.72		65	14.02	12.72	11.60
55	18.64	16.30	14.41		70	13.45	12.21	11.16
60	18.19	15.92	14.10					
65	17.77	15.56	13.79		55	14.52	13.16	12.00
70	17.40	15.24	13.51		60	13.69	12.48	11.44
				55	65	12.93	11.84	10.89
35	20.10	17.43	15.30		70	12.27	11.26	10.38
40	19.53	17.02	15.00		60	12.72	11.68	10.78
45	18.97	16.60	14.68	60	65	11.81	10.92	10.12
50	18.42	16.17	14.35		70	11.02	10.22	9.51
55	17.89	15.75	14.00	65	65	10.72	9.98	9.33
60	17.39	15.34	13.66		70	9.74	9.12	8.57
65	16.93	14.95	13.33	70	70	8.48	8.02	7.60
70	16.52	14.59	13.02					
40	18.91	16.56	14.66					
45	18.29	16.10	14.31					
50	17.67	15.63	13.94					
55	17.09	15.16	13.56					
60	16.53	14.70	13.19					
65	16.02	14.27	12.82					
70	15.57	13.88	12.47					
45	17.59	15.58	13.91					
50	16.91	15.05	13.50					
55	16.25	14.52	13.08					
60	15.62	14.01	12.65					
65	15.05	13.53	12.24					
70	14.55	13.01	11.85					

OBSER-

APPENDIX.

OBSERVATIONS

ON

TABLES I. and II.

THESE Tables may be met with in most of the books that treat of compound interest and annuities; but there has been, in this work, so much occasion for referring to them, that it was necessary to save the reader the trouble of turning to other books for them.

The 2d, 3d, 4th, &c. numbers in the *Second* Table, are only the *sums* of the first 2, 3, 4, &c. numbers in the *First* Table. This Table, therefore, is the foundation of the *Second*; and, indeed, of all the common tables of compound interest; and, with the help of it, almost all the questions in compound interest may be easily answered.

The following specimen of this may, I think, be of considerable use.

QUESTION I. "To what *sum* or *annuity* " will any given *sum* or *annuity*, now to be " laid up for improvement, at a given rate " of compound interest, increase, in a given " number of years?"

AN-

APPENDIX. 345

ANSWER. Divide the given *sum* or annuity by the value of 1 *l*. payable at the end of the given number of years, and the *quotient* will be the answer.

Example. Let the given sum be 50 *l*. and the given time 18 years. The rate of interest 4 *per cent*.—The present value, at 4 *per cent*. of 1 *l*. payable at the end of 18 years is, by Table I. .4936; and 50 *l*. divided by this value, gives *l*. 101.296, or 101 *l*. 5 *s*. the *sum* to which 50 *l*. will increase in 18 years. In like manner; 2 *l. per annum*, the first payment of which is to be made immediately, will be increased (interest supposed the same) at the end of 18 years, to an *annuity* of *l*.4.05: for 2 *l*. the given annuity, divided by .4936, gives *l*. 4.05, or 4 *l*. 1 *s*.

QUESTION II. "To what *sum* will a given *annuity* amount, in consequence of being forborn and improved, at a given rate of compound interest, for a given number of years?"

ANSWER. From the *increased* annuity, found by the last Question; subtract the *given* annuity; and multiply the *remainder* by the PERPETUITY, and the *product* will be the answer.

Example. 2 *l. per ann.* improved at 4 *per cent.* compound interest, will, by the last Question, increase, in 18 years, to *l*. 4.05 *per ann.* 2 *l*. subtracted from 4.05, leaves 2.05, which,

which, multiplied by 25, the *perpetuity*, gives *l*. 51.25, or 51 *l*. 5 *s*, the *amount* in 18 years. In the same manner it may be found, that 10 *l. per ann.* (interest being the same) will amount, in 41 years, to 998 *l*.

It should be remembered, that the PERPETUITY is 33.33,—28.57,—25,—20,—or 16.666; according as interest is reckoned at 3,—3½,—4,—5 or 6 *per cent:* And that the *annuity* meant in all these Questions is an annuity, the first payment of which is to be made immediately.

QUESTION III. " In what number of
" years will a given *sum* or *annuity* increase
" to another given *sum* or *annuity*, in conse-
" quence of being improved at a given rate
" of interest?"

ANSWER. Divide the *original sum* or *annuity* by the *increased sum* or *annuity*; and look for the *quotient*, or the number nearest to it in Table I; and the number of years corresponding to it will be the answer.

Example. Let the *sum* be 50 *l*. The increased sum *l*. 101.29. The rate of interest, 4 *per cent*. The former sum divided by the latter gives .4936, which stands opposite in the Table to 18 years, or the time in which 50 *l*. will gain the required increase.——In like manner, it may be found, that 18 years is the time in which 2 *l. per ann.* will increase to *l.* 4.05 *per ann.*

QUESTION

APPENDIX. 347

QUESTION IV. " In what time will any
" given *annuity* amount to a *given sum*, in
" consequence of being forborn and im-
" proved, at a given rate of compound in-
" terest?"

ANSWER. Divide the given *sum* to which the annuity must amount by the PERPETUITY. Add the given annuity to the quotient; and by the quotient so increased, divide the given annuity; and this *second quotient*, found in Table I. will shew the answer.

Example. A person owes 1000 *l*. and resolves to appropriate 10 *l. per annum* of his income towards discharging it. In what time will such an appropriation, in consequence of being improved at 4 *per cent.* amount to a sum equal to the debt?——— 1000 *l*, divided by 25 gives 40 *l*. 10 *l.* added to 40 *l*. makes 50 *l*; and 10 *l*. divided by 50 *l*. gives .2000, which in the Table stands opposite to 41 years, the required time.

In the same manner it will appear, that the same annuity, if improved at 5 *per cent.* will amount to 1000 *l*. in 37 years.

QUESTION V. " In what time will a
" *given principal* be annihilated, by taking
" out of it, at the end of a year, a given sum,
" and after that, the same sum annually, to-
" gether with its growing interests?"

AN-

ANSWER. In the same time plainly in which an equal annuity would amount to the *given principal*.

A person, therefore, possess'd of 1000 *l*. capital, bearing interest at 4 *per cent*. would, by Question IV. reduce it to nothing in 41 years, by taking out of it 10 *l*. at the begining of the first year, and as much more every following year, as would be necessary, together with the interest of the remaining capital, to make his annual income constantly 50 *l*.

Remark. The sum to which a *given* annuity will amount in a given time, is the same with the value of an annuity for the given time, equal to the *given* annuity increased by the yearly interest of the amount. That is, 1000 *l*. is the value of 50 *l*. *per ann*. for 41 years at 4 *per cent*: And the same sum is likewise the value of 60 *l*. *per annum*, for 37 years at 5 *per cent*. The reason is plain: 1000 *l*. it has appeared, would, in consequence of being put out to these different rates of interest, be just sufficient to pay the annuities.

I have been the more explicit in these rules, because they point out a very easy method of deducing and examining all I have said, in different parts of this work, and particularly in Chap. III. concerning the increase

crease of money at interest.——I will just mention one instance.

400,000 *l. per annum*, applied in the manner supposed in Questions IV. and V. would annihilate 55 millions, bearing interest at 5 *per cent.* in 42 years.

In 1716, when the *sinking fund* was established, the public debts were near this sum, and bore 5 *per cent.* interest. This fund then, had but 400,000 *l.* of it been inviolably applied to the annihilation of the public debts, would, in 1758, have discharged all the debts contracted *before* 1716.——And it may be further found very easily, by the answer to Question IV. that had it been suffered to go on in its operation, and been applied, *since* 1758, to the redemption of only 3 *per cents* at *par*, it would by this time have discharged 104 millions; and seven years hence 140 millions.——The assertion, therefore, in page 165, is strictly true. But the following proof of that assertion will, perhaps, be more clear and striking.

Suppose an annuity of 400,000 *l*, beginning in 1716, to have been applied UNALIENABLY till 1730, to the annihilation of debts bearing interest at 5 *per cent*; from 1730 to 1748, to the annihilation of debts bearing interest at 4 *per cent.* and from 1748 to 1771, to the annihilation of debts bearing interest at 3 *per cent.* In the first of these periods the annuity would have increased to 800,000 *l.*;

in the *second*, to 1,600,000 *l.*; in the *laſt*, to 3,200,000 *l.*———In the laſt year, therefore, the nation might have been eaſed of above **three millions per annum** in taxes. And, at the ſame time, (ſuppoſing all the ſame meaſures taken in other reſpects) it would have enjoyed the benefit of the greateſt part of that very *ſinking fund* it now has; and no detriment could have ariſen to the public, from any of the applications which have been made of it to current expences.

APPENDIX. 351

DIRECTIONS for finding the VALUES of two JOINT LIVES, and of the LONGEST of two lives; and also, of three JOINT LIVES and the LONGEST of three lives, by Tables VII, XI, XV, and XVI.

IF both the ages are given in the Tables, the value wanted will be found immediately by inspection.

If the ages are not given in the Tables, it will be best to proceed in the following manner.

Suppose the rate of interest 4 *per cent.* and the value desired of two joint lives, whose ages are 40 and 66.——It will appear, from inspecting Table VII. that the value sought would be 6.556, were the age of the elder life 65; and 5.383, were it 70. Since, therefore, it is 66, the value must be the *first* of four arithmetical means between 6.556 and 5.383, or 6.322.——For the same reason, had the ages of the elder been 68, the value would have been the 3d arithmetical mean between 6.556 and 5.383 or 5.854.—In like manner, were the proposed ages 43 and 65, the value would be the third arithmetical mean between 6.556 (the value of two joint lives whose ages are 40 and 65) and 6.425, (the value of two joint lives whose ages are 45 and 65) or 6.478.

Again, let the ages be 43 and 66. That is, let it be supposed, that neither of the proposed ages is given in the Table.

The values corresponding to the ages $\{^{40}_{45}\}$ and $\{^{66}_{66}\}$, are $\{^{6.322}_{6.200}\}$.

The value, therefore, corresponding to the ages 43 and 66, must be the 3d mean between 6.322 and 6.200, or 6.250.——

N. B. The 1st, 2d, 3d, and 4th of four arithmetical means between two numbers are found by subtracting $\frac{1}{5}$, $\frac{2}{5}$, $\frac{3}{5}$, and $\frac{4}{5}$ of the *difference* between the two numbers, from the *greatest* of them.

Thus. The difference between 6.556, and 5.383, is 1.173. One-fifth of this difference is .234; which, subtracted from 6.556, leaves 6.322; the first of 4 means between 6.556 and 5.383.——In like manner; the difference between 6.322 and 6.200 is .122. *One* fifth of this difference is .024; and, therefore, three-fifths of this difference is .072, which, subtracted from 6.322, leaves 6.250, the *third* arithmetical mean between 6.322 and 6.200.

In order to avoid trouble, if the ages are nearly equal, a year or two may be added to the least, and as much subtracted from the greatest; and the value taken by inspection.

APPENDIX. 353

But if one of them much exceeds the other, it will in general be sufficient to take the nearest number in the Table for the lesser.

The mean between the values at 3 *per cent*. and 4 *per cent*. may be taken for the value at 3½ *per cent*. without any error of consequence. And the like may be said of the values at 4½ *per cent*.

The values of the *longest* of two lives is found by subtracting the value of the *joint* lives from the *sum* of the values of the two *single* lives.——Thus, the values of two single lives, whose ages are 25 and 30, are by Table VI. (interest reckoned at 4 *per cent*.) 15.31 and 14.68. The sum of these two values is 29.99; the value of the joint lives is (by Table VII) 11.46; and this value, subtracted from 29.99, gives 18.53, or the value of an annuity on the longest of the two lives.—By this rule, Table XVI. has been calculated; and a demonstration of it may be found in Mr. *Simpson's Doctrine of Annuities and Reversions*, page 20.

The value of *two* joint lives being given, the value of *three* joint lives may be found by the following rule, taken from Mr. *Simpson's Select Exercises*, page 279.

Let A be the youngest, and C the oldest of the three proposed lives. Take the value of the two joint lives B and C, and find the

A a   age

age of a *single* life D of the same value. Then find the value of the *joint* lives A and D, which will be the answer.

Example. Let the three given ages be 25, 30, and 40, and let the rate of interest be 4 *per cent*. Then the value of the two oldest joint lives B and C, will (by Tab. VII.) be 10.428, answering, in Tab. VI. to a single life D of 54 years of age. And the value of the joint lives A and D, which is 8.917 years purchase, will be the value sought.

From the value of *three* joint lives given, the value of the *longest* of *three* lives may be deduced in the following method.—" From
" the sum of the values of all the *single*
" lives, subtract the sum of the values of
" all the *joint* lives, combined two and two.
" Then to the remainder add the value of
" the three joint lives; and this last sum
" will be the value of the *longest* of the three
" lives." See Mr. *Simpson's Doctrine of Annuities*, &c. page 23—or Mr. *Dodson's Mathematical Repository*, Vol. I. page 244.

Example. The sum of the values of three single lives, whose ages are 25, 30, and 40, is (reckoning interest at 4 *per cent*.) 43.202. The value of *two* joint lives, whose ages are 25 and 30, is, 11.468; of *two* joint lives, whose ages are 25 and 40, is 10.655; of *two* joint lives, whose ages are 30 and 40, is 10.428, by Table VII; and the sum of these three

three values is 32.551. This sum subtracted from 43.202 leaves 10.651; which remainder added to 8.917 (the value just found of the three joint lives) gives 19.568, the value of the longest of the three lives.

A SUPPLEMENT,

CONTAINING

ADDITIONAL OBSERVATIONS

AND

TABLES.

SINCE the first publication of this work, I have had the pleasure of reading an ingenious Memoir on the State of Population in the *Pais de Vaud*, a district of the province of *Bern*, in *Switzerland*. The author of this memoir is Mr. *Muret*, the first minister at *Vevey*, a town in that district, and secretary to the Oeconomical Society there. It forms the first part of the *Bern* Observations for the year 1766; and a good abstract of it may be found in the 69th article of a work entitled, *De re Rustica*, or the *Repository*. It contains an account of many facts which appear to me curious and important; and which confirm the observations I have made

made in the firſt and fourth Eſſays in this Treatiſe.—Some of theſe facts I will here beg leave to recite.

In the firſt Eſſay I have aſſerted, that there is a much greater difference between the probabilities of life in *great towns* and in *country pariſhes*, than is commonly ſuſpected; and, as one proof of this, I have obſerved, that tho' in *London* the greateſt part of the natives die under three years of age, in the country the greater part live to marry. Mr. *Muret*'s Obſervations and Tables give a diſtinct demonſtration of this, by ſhewing, that in the province of *Vaud*, the greater part of the inhabitants live many years beyond the age of maturity.—But to be a little more explicit.

The diſtrict of *Vaud*, in *Switzerland*, contains 112,951 inhabitants of all ages; 25,778 *families*; 38,328 married perſons; and the annual medium of *births*, for 10 years before 1766, had been 3155; of *weddings*, 808; of *deaths*, 2504.—It appears, therefore, that the married are very nearly a *third* part of the inhabitants, that the number of perſons to a family is $4\frac{1}{3}$; and that one in 45 of the inhabitants die annually. It may be further learnt, by dividing half the number of the married, by the annual medium of weddings, that the *expectation* of marriage in this country is 23 years and $\frac{1}{2}$; and, from the proportions of the births, weddings, and deaths

deaths (*a*), that the greater part of those who are born live to marry. But of this fact there is, I have juſt intimated, a more particular and diſtinct proof.—From a Table given by Mr. *Muret*, of the rate of human mortality in this country, derived from regiſters kept in 43 pariſhes, of the ages at which the inhabitants die, it appears, that one *half* of all that are born live beyond 41 years of age.—The examination of this Table will, undoubtedly, be a gratification to the reader; and, therefore, I have choſen to make it a part of theſe additions. I have alſo here given the Table referred to, in p. 194 and 268, of the probabilities of life in the pariſh of *Holy-Croſs*, near *Shrewſbury*; and a *third* Table, which I have formed from a regiſter in *Suſmilch*'s works, of the ages at which the inhabitants of a country pariſh in BRANDENBURGH died, during 50 years; or from 1710 to 1759.—I have further thought proper to add, as contraſts to theſe Tables, two Tables exhibiting the probabilities of life at VIENNA and BERLIN.—The following obſervations concerning theſe Tables ſhould be attended to.

The Table for the country of VAUD, tho' it gives the probabilities of life in its firſt ſtages very high; and, at ſome ages, more than double to the probabilities of life in great cities; yet, certainly, gives them too low.

(*a*) See the note, p. 196, &c.

low. For, first, it has just appeared, that in this country the births exceed considerably the deaths. The emigrations, likewise, from it are very numerous, as will be presently observed: And the necessary effect of these two causes is, to make the registers give the number of deaths in the first stages of life, too great in comparison of the deaths in the last stages. A Table formed from such registers must give the probabilities of life too low, according to the observations in the 4th Essay; and, in the present case, they must be given so much too low, as to afford sufficient reason for concluding, that the greater part of the births don't become extinct 'till near the decline of life.

After 40, the probabilities of life in this country decrease very fast; and in old age, they appear to be lower than the probabilities of life in great towns. I have assigned the reason of this fact in page 270, &c. All turned of 65 or 70 in great towns, are a selected body consisting of persons seasoned to their situation, and possessed of constitutions particularly strong; and they may, I think, be not improperly compared to a company of persons on a hazardous journey, who are become a set of picked and hardy travellers, in consequence of having lost all the tender and infirm, and been used to inclement weather and fatigue.—Persons of feeble frames may, with the help of the simple manners and pure

pure air of the country, attain to old age; but in great towns they ſtand no chance for this; the effect of which muſt be that, at the ſame time that greater numbers will attain to old age in the country, they will die off faſter. Thus; in the diſtrict of VAUD, the numbers alive at 75 are above double the numbers alive at the ſame age at BERLIN; but thoſe who attain to that age at BERLIN, have a greater *expectation* of life. The ſame may be obſerved of NORTHAMPTON compared with VIENNA and LONDON.—In ſhort; the truth is, however ſtrange it may ſeem, " that the deſtructive influence of great towns " on life is the very reaſon why old people " live longer in them, than in ſmall towns " and in the country."—Mr. *Muret* has taken notice of this fact; but, ſuppoſing it not general, he aſcribes it to the particular prevalency of drunkenneſs in his country. He had, he ſays, once the curioſity to examine the regiſter of deaths in one town, and to mark thoſe whoſe deaths might be imputed to drunkenneſs, and he found the number ſo great, as to incline him to believe, that hard drinking kills more of mankind than pleuriſies and fevers, and all the moſt malignant diſtempers. This, probably, is very true; but the fact I am conſidering is not owing to it. Drunkenneſs cannot be ſuppoſed to prevail more in the country than in great towns. And it always deſtroys long before old age.

The

The observations now made are applicable to the Table for the country parish in *Brandenburgh*; for it appears from *Susmilch*'s account, that the births there exceed the deaths more than in the country of VAUD; nor is it to be imagined, that there are not likewise many emigrations from it, particularly, to BERLIN and the King of *Prussia*'s armies.

From the Tables for VIENNA and LONDON, compared with the Table for BERLIN, it appears that the last of these towns, tho' much the smallest, has at some ages even a worse effect on the duration of life than either of the former: And the reason, perhaps, may be, that the inhabitants there are much more crouded together. See p. 225.—Between the ages of 30 and 35, and also between 42 and 52, there is an irregularity in the BERLIN Table, which, very probably, would not have appeared in it, had it been formed from the bills for a longer term of years.—The like observation might be made on an irregularity in the 2d Table, between the ages of 25 and 30.

From the age of 25 to 45, VIENNA appears, in the Tables, to be less unfavourable to life than LONDON; but it cannot be depended upon that this is the truth, for the VIENNA Table may give the probabilities of life at these ages higher, only because the recruits from the country come to it later, or in greater

greater numbers, after 30 and 40, than in LONDON. A like effect would also arise from a greater number of migrations in old age from LONDON than from VIENNA. See the note, p.

In forming the Tables for VIENNA and BERLIN, I have applied the correction explained in the 4th Essay, and demonstrated there to be necessary; and, in making this correction, I have supposed, agreeably to the proportion of the births to the burials, that a fifth of all who die in these cities, are persons who removed to them at 20 years of age.—Notwithstanding this correction, the Table for BERLIN gives the probabilities of life between 10 and 20 so high, and in such disproportion to the probabilities of life immediately after 20, as to exceed all the bounds of credibility. The true reason of this may be learnt from what has been said in p. 225, of the rapid increase of BERLIN.

My chief purpose in giving these Tables is to exhibit, in the most striking light, the *difference* between the state and duration of human life, in *great cities* and in the *country*. It is not possible to make the comparison, without concern and surprize. I will here beg leave to lay it in one view before the reader, desiring him to take with him this consideration, that, for the reasons I have explained, it can be erroneous only by giving the difference (*a*) much too *little*.

(*a*) See p. 222, &c. p. 252, p. 246.

Pro-

Proportion of Inhabitants dying annually in

Pais De Vaud	Country Parish in Brandenburg	Holy-Crofs near Shrewfbury	London	Vienna	Berlin.
1 in 45	1 in 45	1 in 33	1 in $20\frac{3}{4}$	1 in $19\frac{1}{2}$	1 in $26\frac{1}{2}$ (a)

Ages to which half the born live.

Pais De Vaud	Country Parish in Brandenburg	Holy-Crofs	London	Vienna	Berlin.
41	$25\frac{1}{2}$	27	$2\frac{3}{4}$	2	$2\frac{3}{4}$

Proportion of the Inhabitants (b) who reach 80 years of Age.

Pais De Vaud	Country Parish, Brandenburgh	Holy-Crofs	London	Vienna	Berlin.
1 in $21\frac{1}{2}$	1 in $22\frac{1}{2}$	1 in 11	1 in 40	1 in 41	1 in 37

The

(a) See page 225. This proportion, were there either no increafe, or but a flow increafe at BERLIN, would certainly be found to be much the fame with that in VIENNA and LONDON.

(b) It fhould be recollected here, that a confiderable part of thofe who die turned of 80 years of age in great towns, are *emigrants* from the country, who came to them in full maturity, after efcaping the weaknefs of infancy. And that alfo in general thefe *emigrants* confift of the more hearty and robuft part of the kingdom. On both thefe

The (a) Probabilities of living one Year in

Odds	Pais De Vaud	Country Parish, Brandenburgh	Holy-Cross	London	Vienna	Berlin
At birth	$4\frac{1}{4}$ to 1	$3\frac{1}{2}$ to 1	$4\frac{1}{2}$ to 1	2 to 1	$1\frac{1}{3}$ to 1	$1\frac{1}{4}$ to 1
Age 12	160 to 1	112 to 1	144 to 1	75 to 1	84 to 1	123 to 1
25	117 to 1	110 to 1	100 to 1	56 to 1	66 to 1	50 to 1
30	111 to 1	107 to 1	96 to 1	45 to 1	56 to 1	44 to 1
40	83 to 1	78 to 1	55 to 1	31 to 1	36 to 1	32 to 1
50	49 to 1	50 to 1	50 to 1	24 to 1	27 to 1	30 to 1
60	23 to 1	25 to 1	26 to 1	18 to 1	19 to 1	18 to 1
70	$9\frac{1}{2}$ to 1	11 to 1	16 to 1	12 to 1	11 to 1	12 to 1
80	4 to 1	6 to 1	8 to 1	7 to 1	7 to 1	7 to 1

EXPECTATIONS of Life.

	Pais De Vaud	Country Parish in Brandenburgh	Holy-Cross	London	Vienna	Berlin
At birth	37 yrs	$32\frac{1}{2}$ years	$33\frac{1}{4}$ yrs	18 yrs	$16\frac{1}{2}$ yrs	18 yrs
Age 12	$44\frac{3}{8}$	44	$43\frac{1}{2}$	$33\frac{1}{2}$	$35\frac{1}{4}$	$35\frac{1}{2}$
25	$34\frac{1}{4}$	$35\frac{1}{2}$	35	26	$28\frac{1}{4}$	$27\frac{3}{4}$
30	$31\frac{1}{4}$	$31\frac{1}{2}$	32	$23\frac{1}{2}$	$25\frac{1}{2}$	$25\frac{1}{4}$
35	$27\frac{1}{2}$	28	$28\frac{1}{2}$	$21\frac{1}{2}$	$22\frac{1}{2}$	$22\frac{1}{2}$
40	24	25	$25\frac{1}{2}$	$19\frac{1}{2}$	$20\frac{1}{2}$	$20\frac{1}{2}$
45	$20\frac{1}{2}$	$21\frac{1}{2}$	$23\frac{1}{4}$	$17\frac{1}{2}$	$17\frac{1}{2}$	$18\frac{1}{2}$
50	$17\frac{1}{2}$	18	20	16	16	$16\frac{1}{4}$
55	$14\frac{1}{2}$	15	17	$14\frac{1}{4}$	$13\frac{1}{2}$	14
60	12	$12\frac{1}{2}$	$14\frac{1}{2}$	$12\frac{1}{2}$	$11\frac{1}{2}$	$12\frac{1}{2}$
65	$9\frac{1}{2}$	$9\frac{1}{2}$	$11\frac{1}{4}$	$10\frac{1}{2}$	$9\frac{1}{2}$	$10\frac{1}{2}$
70	$7\frac{1}{2}$	$7\frac{1}{2}$	10	$8\frac{1}{4}$	$8\frac{1}{2}$	$8\frac{1}{4}$
75	$5\frac{1}{2}$	$5\frac{1}{2}$	8	7	$6\frac{1}{2}$	7
80	$4\frac{1}{2}$	$4\frac{1}{2}$	5	5	$5\frac{1}{2}$	6

From these accounts the numbers attaining to old age in great towns ought to be much greater than in the country. In *London*, *Vienna*, and *Berlin*, they ought to be nearly *double*; but we see, that, in reality, they are scarcely *half*.

(*a*) These probabilities are here given sufficiently near for the present purpose, and so as to err on the side favourable

From this comparison it appears with how much truth great cities have been called the *graves* of mankind. It must also convince all who will consider it, that, according to the observation at the end of the 4th Essay, it is by no means strictly proper to consider our diseases as the original intention of nature. They are, without doubt, in general, our own creation. Were there a country, where the inhabitants led lives entirely natural and virtuous, few of them would die without measuring out the whole period of present existence allotted them; pain and distempers would be unknown among them; and the dismission of death would come upon them like a sleep, in consequence of no other cause than gradual and unavoidable decay.— Let us then, instead of charging our Maker with our miseries, learn more to accuse and reproach *ourselves*.

The reasons of the baleful influence of great towns, as it has been now exhibited, are plainly,

First, The irregular modes of life, the luxuries, debaucheries, and pernicious customs, which prevail more in towns than in the country.

vourable to towns; but the manner of forming the Tables is such, that they sometimes give them irregularly, and always with less correctness than the *expectations*, or the same probabilities for *periods* of years.

Secondly,

Secondly, The foulness of the air in towns, occasioned by uncleanliness, smoak, the perspiration and breath of the inhabitants, and putrid steams from drains, church-yards, kennels, and common-sewers.—It is, in particular, well known that air, spoiled by breathing, is rendered so noxious, as to kill instantaneously, any animal that is put into it. There must be causes in nature (*a*) continually operating, which restore the air after being thus spoiled. But in towns it is, probably, consumed faster than it can be adequately restored; and the larger the town is, or the more the inhabitants are crouded together, the more this inconvenience must take place.

But I must proceed to some more of Mr. *Muret*'s observations.———In the 4th Essay, p. 271, &c. I have given an account of several facts which prove the probabilities of life to be higher among females than males. Agreeably to this it appears, that in the dis-

---

(*a*) A well-known and excellent philosopher has for some time been employed in enquiring into these causes; and he has made several curious and important discoveries, of which I hope the world will soon receive a particular account. One of these discoveries has been lately published in a pamphlet, entitled, *Directions for impregnating Water with fixed Air, in order to communicate to it the peculiar Spirit and Virtues of Pyrmont Water, and other Mineral Waters of a similar Nature.* By the Rev. Dr. PRIESTLY.

trict

trict of VAUD, half the *females* don't die till the age of 46 and upwards, tho' half the *males* die under 36. This great difference is in some measure owing to the military and commercial emigrations among the males; but it appears undeniably, that their greater mortality contributes likewise to it. The number of *males* who died, for a course of years, in 39 parishes of this district, was 8170; of *females* 8167; of whom the numbers that died under one year of age were 1817 *males*; and 1305 *females*; and under 10 years of age; 3099 *males*, and 2598 *females*. In the *beginning* of life, therefore, and before any emigrations can take place, the rate of mortality among *males* appears to be much greater than among *females*: And this is rendered yet more certain, by the account Mr. *Muret* gives of the proportions of the deaths among males and females in the *first* year of life at VEVEY. In this town, he acquaints us, that for 20 years ending in 1764, there died in the first month, of *males* 135, to 89 *females*; and, in the first year, 225 to 162.——To the same effect it appears, from a Table given by *Susmilch* (*a*), that in BERLIN 203 *males* die in the first month, and but 168 *females*; and in the first year, 489 to 395; and also, from a Table of *Struyck*'s, that in HOLLAND, 396 *males* die in the first *year*, to 306 *females*.—What is

---

(*a*) See *Susmilch's Gottliche Ordnung*, Vol. II. p. 317, &c.

moft of all remarkable is, that thefe accounts fhew, that both at VEVEY and BERLIN the *ftill-born males* are to the *ftill-born females*, as 30 to 21, or nearly in the proportion given by the accounts referred to in p. 274.

The whole number of inhabitants at VEVEY in 1764, was 3350. Of thefe 1931 were females, and only 1419 males. Sixty-fix were *widowers*, and 200 *widows*. The number of *batchelors*, above 16 years of age, was 529; and of *virgins*, above 14 years of age, 734. See Mr. *Muret*'s Tables, p. 124.

Mr. *Deparcieux* at PARIS, and Mr. *Wargentin* in SWEDEN, have obferved, that not only *women* live longer than *men*, but that *married* women live longer than *fingle* women. The regifters examined by Mr. *Muret* confirm this; and it appears particularly, that, of equal numbers of *fingle* and *married* women between 15 and 25, more of the former died than of the latter, in the proportion of 2 to 1. The reafon of this may be, as Mr. *Muret* acknowledges, that the women who marry, are a felected body, confifting of the more healthy and vigorous part of the fex. But this, probably, is by no means the only reafon; for it may, I think, be expected, that in this, as well as in all other inftances, the confequences of following nature muft be favourable.

The facts recited here, and at the end of the 4th Effay, prove, beyond the poffibility of denial,

B b

denial (*a*), that there is a difference between the mortality of males and females.—I must however observe, that it may be doubted, whether this difference, so unfavourable to males, is *natural*; and the following facts will prove, that I have reason for such a doubt.

It appears, from several registers in *Susmilch*'s works, that this difference is much less in the *country parishes* and *villages* of BRANDENBURGH, than in the *towns*: And, agreeably to this, it appears likewise, from the accounts of the same curious writer, that the number of males in the country comes much nearer to the number of females.

In 1056 small *villages* in BRANDENBURGH, the *males* and *females*, in 1748, were 106,234, and 107,540, or to one another as 100 to $101\frac{1}{3}$. In twenty small *towns* they were 9544, and 10,333; or as 100 to $108\frac{1}{4}$. In BERLIN they were, exclusive of the garrison, 39,116 and 45938; or as 100 to $117\frac{1}{2}$.

At the time the accounts, mentioned in p. 206, were taken of the inhabitants in the

---

(*a*) In the printed ACCOUNT of the Society in *Nicolas-Lane*, for *Equitable Assurances on Lives and Survivorships*, there is a Table of the values of assurances on *female lives*, which supposes them to be more hazardous than *male lives*. This Table is derived from an opinion generally received at the time it was composed; but I am desired to inform the public, that no such Table shall be admitted into the future editions of that ACCOUNT; the society being determined to maintain the just credit it has acquired, by keeping strictly, in every instance, to calculations, founded on the best observations.

province of New Jersey in America, they were diſtinguiſhed particularly into *males* and *females* under and above 16.

In 1738, the number of
Males under 16 was, 10639. Females 9700
Males above 16 ——— 11631. Females 10725

In 1745, theſe numbers were,
Males under 16 — 14523. Females 13754
Males above 16 — 15087. Females 13704

The inference from theſe facts is very obvious. They ſeem to ſhew ſufficiently, that human life in males is more brittle than in females, only in conſequence of adventitious cauſes, or of ſome particular debility, that takes place in poliſhed and luxurious ſocieties, and eſpecially in great towns (*a*).

From the proportion of the births to the deaths in the diſtrict of Vaud, as mentioned in p. 358, it follows, by the rule in the note p. 208, that the inhabitants ought to double their

---

(*a*) The number of deaths for 60 years at Vevey, in the four *winter* months, (December, January, February and March) were to the deaths in the four *ſummer* months (June, July, Auguſt, and September) as 2140 to 1697, or 5 to 4. (See Mr. *Muret*'s Tables, p. 100). In London and at Paris, this proportion is nearly the ſame. At Edinburgh, as 4 to 3. In 25 country towns and pariſhes mentioned by Dr. *Short* (*New Obſervations*, p. 142) as 50 to 41.—The ſick admitted into the *Hotel Dieu* at *Paris*, for 40 years, from 1724 to 1763, were,

their own number in 120 years. But the fact is, that so many migrate into foreign armies and with commercial views, that their increase is scarcely sensible. Mr. *Muret*, after observing this, enters into a general account of the causes which obstruct population in his country. Among these he insists particularly on LUXURY and the ENGROSSING OF FARMS. I wish his observations on these subjects were not applicable to the present state of this kingdom: But, perhaps, there is no kingdom in the world to which they are *so* applicable.—In consequence of the easy communication lately created, between the different parts of the kingdom, the LONDON fashions and manners, and pleasures, have been propagated every where; and almost every distant town and village now vies with the capital in all kinds of expensive dissipation and amusement. This enervates and debilitates; and, together with our taxes, raises every where (*a*) the price of the

---

were, in the former months, 314,824; in the latter, 238,522, or as 4 to 3. See *Recherche's sur la Population*, &c. per *M. Messance*, p. 181. And agreeably to all this, Dr. *Percival* informs me, that at *Manchester* the mortality of *winter* and *summer* are to one another as 11 to 8.—It is remarkable that the *births* also in *winter* to those in *summer* are at VEVEY as 5 to 4; in LONDON as 8 to 7; in the country towns and parishes just mentioned, as 7 to 6.

(*a*) The price of corn, in particular, has for some time been complained of by the poor as oppressively high, though

the means of subsistence, checks marriage, and brings on poverty, dependance, and venality.—With respect, particularly, to the custom of *engrossing farms*, Mr. Muret observes, with the highest reason, that a large tract of land, in the hands of one man, does not yield so great a return, as when in the hands of several, nor does it employ so many people; and, as a proof of this, he mentions two parishes in the district of VAUD, one of which (once a little village) having been bought by some rich men, was sunk into a single *demesne*; and the other, (once a single *demesne*) having fallen into the hands of some peasants, was become a little village.—How many facts of the former kind can this country now furnish?—And there is reason to apprehend they will go on increasing.—The custom of engrossing farms eases *landlords* of the trouble attending the necessities of little tenants and the repairs of cottages.—A great farmer, by having it more in his power to speculate and to command the markets, and by drawing to himself the profits which would have supported several farmers, is capable, with less culture, of paying a higher rent.

though far from being so high as it generally was at the end of the last century. This is a striking fact which implies that the *lower* part of the nation are now more distressed than ever. The consequence has been a reduction of their number; and this is an effect that must go on increasing, with increasing luxury and taxes.

Our superiors, therefore, find their account in this evil.—But it is, indeed, erecting *private* benefit on *public* calamity; and, for the sake of a temporary advantage, giving up the nation to depopulation and distress.——We have, for many years, been feeling the truth of this observation.

Dr. *Davenant*, (the best of all political writers), tells us, that at *Michaelmas*, in the year 1685, it appeared by a survey of the hearth-books (*a*) that the number of houses in all ENGLAND and WALES was 1,300,000, of which 554,631 were houses of only one chimney. See Dr. *Davenant*'s Works, Vol. II. p. 203.—In his *Essay on Ways and Means*, &c. Vol. I. p. 33, he gives a particular account of the number of houses in every county, according to the *hearth-books* of Lady-day, 1690; and the sum total then was 1,319,215. —At the *restoration* it appeared by the same hearth-books, that the number of houses in the kingdom (*b*), was 1,230,000.—In the

---

(*a*) At this time there was a tax of two shillings on every *fire-hearth*; which was taken off at the REVOLUTION, because reckoned "not only a great oppression to the "poorer sort, but a badge of slavery on the whole peo- "ple, exposing every man's house to be entered into "and searched at pleasure by persons unknown to him." *Preamble to the act for taking away the revenue arising by hearth-money*, 1 William *and* Mary, Chap. 10.

(*b*) Continuation of *Rapin*, Vol. I. p. 53.

interval,

interval, therefore, between the *restoration* and the *revolution*, the people of ENGLAND had increased above 300,000; and " of " SMALLER TENEMENTS, Dr. *Davenant* " observes (*a*), there had been, from 1666 to " 1688, about 70,000 new foundations laid."
—But what a melancholy reverse has taken place since?—In 1759 the number of houses in ENGLAND and WALES was 986,482; of which not more than 330,000 were houses having less than seven windows; and 282,429 were *cottages* not charged on account of poverty.——In 1766, notwithstanding the increase of buildings in LONDON, the number of houses was reduced to 980,692 (*b*); of which 276,149 were *cottages* not charged. According to these accounts then, our people have, since the year 1690, decreased

near

(*a*) Dr. *Davenant*'s Works, Vol. I. p. 370.

(*b*) See *Considerations on the Trade and Finances of this Kingdom*, p. 95, 97, 98. Printed for *Wilkie*, 1766. See also p. 184, &c. of this Treatise; and my *Appeal to the Public on the Subject of the National Debt*, p. 86, &c.—It deserves particular notice, with respect to the accounts here given of the number of houses in 1759 and 1766, that, being returns made by the surveyors of the house and window-duties throughout all ENGLAND and WALES, they are subject to no such deficiencies as those in the account of the number of houses in LONDON, taken by Mr. *Maitland* from the *parish* books, and mentioned in the note, p. 182.—The reason is, that no landlord or tenant can ever consent that any *two or more houses* belonging to him, should be charged by the assessors of the window-tax as

single

near a *million and a half*.—And the waste has fallen principally on the inhabitants of cottages; nor indeed could it fall any where more unhappily; for, from cottages our navies and armies are supplied, and the lower people are the chief strength and security of every state.—What renders this calamity more alarming is, that the inhabitants of the cottages thrown down in the country, fly to LONDON and other towns, there to be corrupted and perish (*a*),—I know I shall be here told that the *Revenue* thrives. But this is not a circumstance from which any encouragement can be drawn. It thrives, by a cause

single houses; because, in this case, he would be taxed too high, and pay more than the law required.——For instance. A building having 20 windows, divided into two distinct tenements, with a family in each, if charged as a *single* house, would pay, besides 3 *s.* for the house, 1 *s.* 7 *d.* for every window, or 1 *l.* 13 *s.* 10 *d.* in all: whereas, if reckoned what it really was, two contiguous houses, it would pay, supposing 10 windows in each tenement, 6 *s.* to the house duty, and only 10 *d.* for each window, or 1 *l.* 2 *s.* 8 *d.* in all.—The number of houses, therefore, subject to the house and window-duty, given in the above returns, must probably be the full number of such houses in the kingdom.

(*a*) Dr. *Davenant* says, from Mr. *King*'s Observations, " that the supply of LONDON alone takes up above *half* " the neat increase of the kingdom."—Is it then to be wondered at, that the supply of the waste in *all* the towns of the kingdom, added to that increase of luxury and taxes, and of the drain to our *armies*, and *navies*, and *foreign settlements*, which has taken place within these 70 years, should have so far exceeded the increase of the kingdom, a cause that is likely in time to destroy both itself and the kingdom; I mean, by an increase of luxury, producing such an increase of *consumption* and *importation* (*a*), as secretly *accelerates* ruin, while *at present* (as far as the Revenue is concerned) it overbalances the effects of depopulation.——What remedies can be applied in such circumstances?—This is a question of great importance, which requires a more deep and careful discussion

as to produce the depopulation I have mentioned?—It has been asserted by political calculators, that no population can bear more than one soldier for every hundred souls. This is saying a great deal too much; but were it true, the number of our soldiers and sailors, even in *peace*, would alone be sufficient to reduce us to nothing in a little time.

A flourishing commerce, tho' favourable to population in some respects, is, I think, on the whole, extremely unfavourable; and, while it flatters, may be destroying: particularly, by increasing luxury, the worst enemy of population as well as of public virtue; and by calling off too many persons from agriculture to unhealthy trades and the sea-service.—Suppose 50,000 sailors, added to other burdens, to have been formerly the whole number the nation could bear without decreasing. In such circumstances, it is plain, that any causes which doubled or tripled that number, would depopulate with rapidity.

(*a*) For Example. In LONDON, those who used to satisfy themselves with *one* house, or perhaps *half* a house, must now have *two* houses. Those who used to live plain must now live high; and those who used to *walk*, must now be *carried*. This is the reason of the increase of consumption and of buildings in LONDON, and not an increase of the inhabitants, for the number of inhabitants is certainly less now than it was forty years ago. Vid. page 190.

than

than I am capable of giving it. I will, therefore, only anfwer generally and *briefly* in a ftyle and language fimilar to Mr. *Muret*'s.

Enter immediately into a decifive enquiry into the ftate of population in the kingdom.—Promote agriculture.—Drive back the inhabitants of towns into the country.—Eftablifh fome regulations for preferving the lives of infants.—Difcourage luxury, and celibacy, and the ingroffing of farms.—Let there be entire liberty; and maintain public peace by a government founded not in *conftraint*, but in the *refpect* and the *hearts* of the people.—But above all things, if it be not now too late; " find out means of avoiding the mife-
" ries of an impending bankruptcy, and of
" eafing the nation of that burden of debts
" and taxes under which it is finking."

POSTSCRIPT.

Containing an Account of the Influence of the different States of civil Society on Population; of the Policy of former Times with respect to Inclosures, engrossing of Farms, and the Encouragement of Agriculture; and also of the State of the lower Classes of Men formerly, compared with their State at present.

THE following observations and facts have lately occurred to me in reconsidering the present state of population in this kingdom; and as, perhaps, they are of some importance, I shall beg leave to introduce them in this place.

One of the most obvious divisions of the state of mankind is, into the *wild* and the *civilized* state. In the former, man is a creature rude, ignorant, and savage; running about in the woods; and living by hunting, or on the spontaneous productions of the earth. In this state, the means of subsistence being scarce, and a large quantity of ground necessary to support a few, there can never be any considerable increase.—In the latter state, man is a creature fixed on one spot, employing himself in cultivating the ground, and enjoying the advantages of science, arts, and civil government. Of this last state there are many different degrees or stages, from the most simple to the most refined and luxurious. The first or the simple stages of civilization, are those which favour most the increase and the happiness of mankind: For in these states, agriculture supplies plenty of the means of subsistence; the blessings of a natural and simple life are enjoyed; property is equally divided; the wants of men are few, and soon satisfied; and families are easily provided for.——On the contrary. In the refined states of civilization property is engrossed, and the natural equality of men subverted; artificial necessaries without number are created; great towns propagate contagion and licentiousness; luxury and vice prevail; and, together with them, disease, poverty, venality, and oppression. And there is a limit at which, when the corruptions of civil society arrive, all liberty, virtue, and happiness must be lost, and complete ruin follow.——Our *American* colonies are at present, for the most part, in the first and the happiest of the states I have described; and they afford a very striking proof of the effects of the different stages of civilization on population. In the inland parts of NORTH-AMERICA, or the back settlements, where the modes of living are most simple, and almost

most every one occupies land for himself, there is an increase so rapid as to have hardly any parallel. Along the sea-coast, where trade has begun to introduce refinement and luxury, the inhabitants increase more slowly: And in the maritime towns (if I may judge from the bills of mortality at BOSTON, mentioned in page 200) they do not increase at all (*a*).

But to confine my thoughts to my own country.—Here, it is too evident that we are far advanced into that last and worst state of society, in which false refinement and luxury multiply wants, and debauch, enslave, and depopulate.—Among the evils of this state, and the causes of depopulation, I have mentioned the accumulation of property. As this is an evil which has been for some time increasing among us, I will give a brief account of its tendencies and effects, with a view, particularly, to the present circumstances of this kingdom, and to some objections which have been started.

By the laws of *Licinius*, no Roman was to hold more than seven *jugera* of land. " Only
" revive, says Mr. *Sufmilch*, this law, or
" that of *Romulus*, which limited every Ro-
" man to two *jugera*, and you will soon

(*a*) Along the sea-coast they double their own number in about 35 years; but in the back-settlements, in 15 years. See *Essay* I. page 206; and *A Discourse on Christian Union*, by Dr. STYLES, p. 109.

" convert

"convert a barren defart into a bufy and crouded hive."—The doubts of fome ingenious men on this fubject, have, indeed, greatly furprized me. I can fcarcely think of a more evident maxim, than that " the divifion of property promotes population."—Let a tract of ground be fuppofed in the hands of a multitude of little proprietors and tenants, who maintain themfelves and families by the produce of the ground they occupy, by fheep kept on a common, by poultry, hogs, &c.; and who, therefore, have little occafion to purchafe any of the means of fubfiftence. If this land gets into the hands of a few great farmers, the confequence muft be, that the little farmers will be converted into a body of men who earn their fubfiftence by working for others, and who will be under a neceffity of going to market for all they want. And, fubfiftence in this way being difficult, families of children will become burdens, marriage will be avoided, and population will decline.—— At the fame time there will, perhaps, be more labour, becaufe there will be more compulfion to it. More bread will be confumed, and, therefore, more corn grown; becaufe there will be lefs ability of going to the price of other food. Parifhes, likewife, will be more loaded, becaufe the number of poor will be greater. And towns and manufactures will increafe, becaufe more will be be driven to them in queſt of places and employments.—This is the way in which the engroſſing of farms naturally operates: And this is the way in which, for many years, it has been actually operating in this kingdom.

It deſerves particular notice, that the obſervations now ſuggeſted ſhew, that the very cauſes which produce depopulation among us, may, for ſome time, promote tillage; and I will take this opportunity to add, that they will alſo account for the following fact.—In the year 1697, wheat was at 3 *l.* a quarter, and other grain proportionably dear. But there was no clamour, and the exportation went on. See a valuable and uſeful Pamphlet, entitled, *Three Tracts on the Corn Trade,* page 100, 107, 145. At preſent, though the quantity of money in the kingdom is doubled; when wheat is at 2 *l.* 8 *s.* a quarter, and in general before any grain, except oats, gets above the prices at which the law allows a bounty on exportation, there is an alarm, the poor are ſtarving, inſurrections begin, and the exportation is prohibited.—I referred to this fact in the note, p. 372; and the true reaſon of it ſeems to be, that the high price of bread was not, at the time I have mentioned, of eſſential conſequence to the lower people, becauſe they could live more upon other food which was then cheap; and becauſe alſo being

more

more generally occupiers of land, they were less under a necessity of purchasing bread. Whereas now, being forced by greater difficulties, and the high price of all other food, to live principally or solely on bread, if that is not cheap, they are rendered incapable of maintaining themselves.

In confirmation of this account, I will beg leave to mention, that, though during the whole last century, corn (wheat, rye, oats, and barley) was generally dearer than it has been, at an average, for the last 40 years; yet flesh-meat was about half its present price: And that, in an *Act of Parliament* of the 25th of Henry VIII. beef, veal, pork, and mutton are mentioned as the food of the poor, and their price limited to about a half-penny a pound. See Mr. *Hume's History of the Tudors*, Vol. II. page 285. Beef and pork, in particular, were sold in LONDON at two pounds and a half, and three pounds for a penny; at the same time that wheat was at 7 *s.* and 8 *s.* a quarter (*a*), and bore the

(*a*) Even so far back as the year 1463, the price of wheat was reckoned not too high at 6 *s.* 8 *d.* per quarter; nor that of barley at 3 *s.* and rye at 4 *s.*; for it was in that year enacted, that the *importation* of these three sorts of grain should not be allowed till they got above these prices. See Mr. *Anderson's Chronological Deduction of Commerce*, Vol. I. page 280.

By a Statute of 1 *Philip* and *Mary*, 1553, leave was given to *export* these three kinds of grain till they rose to these prices. *Ib.* p. 387.

By

the same proportion to the price of flesh as it would bear now, were it at about 4 *l.* a quarter.

By an ordinance in 1563, the exportation prices were fixed to 10 *s.* per quarter for *wheat*; 8 *s.* for *rye, pease,* and *beans*; and 6 *s.* 8 *d.* for *malt*.—And in 1593, to 1 *l.* for *wheat*; 13 *s.* 4 *d. pease* and *beans*; and 12 *s. barley* and *malt.* Ib. p. 401 and 442.

PRICES per QUARTER,

		Of Wheat.			Of Malt.			Of Oats.		
		l.	*s.*	*d.*	*l.*	*s.*	*d.*	*l.*	*s.*	*d.*
In	1491,	0	14	8	0	0	0	0	00	0
	1494,	0	4	0	0	0	0	0	00	0
	1504,	0	5	8	0	0	0	0	00	0
	1512,	0	6	2	0	4	0	0	2	0
	1521,	1	0	0	0	0	0	0	00	0
From 1553 to —	1556,	0	8	0	0	5	0	0	00	0
Before harvest, in	1557,	2	13	4	2	4	0	0	00	0
After harvest, in	1557,	0	8	0	0	5	0	0	10	0
	1560,	0	8	0	0	5	0	0	5	0
Before harvest, in	1574,	2	16	0	0	0	0	0	00	0
After harvest, in	1574,	1	4	0	0	0	0	0	00	0
	1587,	3	4	0	0	0	0	0	00	0
A dearth occasioned by excessive exportation; & in 1596 by great rains	1594,	2	16	0	0	0	0	0	00	0
	1595,	2	13	4	1	0	0	0	00	0
	1596,	4	0	0	1	6	8	0	00	0
	1597,	5	4	0	2	6	4	0	00	0

AVERAGE PRICE,

From 1606 to —	1706,	1	18	6	1	2	0	0	00	0
From 1707 to —	1765,	1	12	6	1	1	9	0	00	0
From 1766 to —	1772,	2	3	6	0	0	0	0	19	0

See Bp. *Fleetwood's Chronicon Pretiosum*, from p. 113 to p. 124. And *Three Tracts on the Corn Trade*, p. 98, &c.

quarter. See *Chronicon Pretiosum*, p. 116.— It appears, indeed, that our ancestors took great

With these prices of *corn* let us compare the prices of *flesh*, at two or three different periods.

In 1512, the price of wheat was from 5 s. 8 d. to 6 s. 8 d. in *Yorkshire*. See the *Regulations and Establishment of the Houshold of Henry Algernon Percy, the fifth Earl of Northumberland, at his Castles of Wresill and Lekingfield, in Yorkshire*, begun Anno Dom. 1512, page 2, 4. Let us call the mean price 6 s. 2 d. The price of malt was 4 s. and of oats, 2 s. We may therefore reckon, that the *nominal* price of grain at this time was about a seventh of its *nominal* price for the last 20 years.

The price of a fat ox at the same time, and in the same county, was 13 s. 4 d; of a lean ox, 8 s; of a weather, 1 s. 8 d; of a calf, 1 s. 8 d; of a hog, 2 s. Ib. p. 5. 6, 7.—The nominal price of meat, therefore, was no more than about a 15th of its present price, and bore the same proportion to the price of corn that it would now bear, were it at *half* its present price.—A like inference may be drawn from comparing the following prices;

Wheat, in 1549, was about 12 s. per quarter in London. Malt, 10 s. Barley, 9 s. Rye, 6 s. 6 d, Oats, 4 s.—A middling ox, 1 l. 18 s. A weather, 3 s. Butter, *three farthings* and a *penny* a pound. Cheese, a *halfpenny* a pound. See *Maitland*'s History of London, page 143, 144.

" In 1574, there was a great dearth, and wheat was, " before harvest, at 2 l. 16 s. per quarter; and beef at " *Lammas* so dear, as to be sold at twopence-halfpenny " a pound." See *Chronicon Pretiosum*, p. 123. That is, beef compared with wheat, was at least one half cheaper than it is now.

In 1445, wheat was at 4 s. 6 d. per quarter. In 1447, at 8 s. In 1448, at 6 s. 8 d. In 1449, 5 s.—A bullock, in 1445, 5 s. A sheep, 2 s. 5 d.$\frac{1}{2}$ A hog, 1 s. 11 d.$\frac{1}{2}$
——Fine cloth for surplices, in 1446, 8 d. per ell. Cloathing for a year, at the same period, of a common servant,

great care to keep the price of flesh low for the poor; and this was one of the reasons of the many proclamations published by Queen *Elizabeth*, *James* I. and *Charles* I. against eating flesh in Lent and on fish days; and against the erection of new buildings in *London*, and the residence in it of the nobility and gentry.

The reason now assigned accounts farther for the great variations in the price of grain which used to take place formerly. These were such as could not be now endured; but, bread being then less a necessary article of subsistence, they were less felt and regarded.

I have taken for granted, in these observations, that the quantity of ground brought

---

servant of husbandry, 3 s. 4 d. Of a chief carter and shepherd, 4 s. Of a bailiff of husbandry, 5 s. Ib. page 198, 109, 160.—*Cloathing*, therefore, at this time, seems to have been cheaper in comparison of the price of corn than even flesh."

"The weight of silver coin *formerly*, to the weight of silver coin of the same denomination *now*, was from 1461 to 1509, as 62 to 37½. From 1509 to 1543, as 62 to 45. From 1552 to 1600, as 62 to 60. And from 1600 to the present time, as 62 to 62." But nothing depends on this in the present enquiry; the object of which is, not the proportion of the prices of the different articles of subsistence *now* to their prices *formerly*, but the proportion TO ONE ANOTHER of their prices *now*, in comparison with the same proportion *formerly*. And this may be as well deduced from the *nominal* as from the *absolute* prices.—Thus, The price of bread now is nearly the same that it was 100 years ago; but, in *comparison* with the price of beef and mutton, it is at least *one half* cheaper.

under tillage in this kingdom is now more than ever it was. This is generally believed; and, if true, the causes of it have been those I have mentioned, in conjunction with the encouragement given to the growth of corn by the bounty on exportation, and the increase of luxury occasioning an increase of horses, and rendering even the poor averse to all bread except that made of the (*a*) finest flour. But, perhaps, the fact may not be so certain as some think it. At least, there is reason to apprehend, that whatever the increase of tillage might have been for 50 or 60 years after the Revolution, it is now at an end.—I have lately received an account of a large common field in *Leicestershire*, which used to produce annually 800 quarters of corn, besides maintaining 200 cattle; but which now, in consequence of being inclosed and getting into few hands, produces little or no corn; and maintains no more cattle than before, though the rents are considerably advanced.—This is only one instance among many of an evil that has been prevailing for some time, and which is the general effect of the laws for inclosing open

---

(*a*) Bread made of *bran*, and even of *pease* and *beans*, was formerly not uncommon among the lower people. But no distresses could force them now to eat such bread, or even to live upon *rice*, though the food of a considerable part of the rest of mankind. See the *Earl of Northumberland's Houshold Book*, Preface, p. 13, &c.

fields.—

fields.—In *Northamptonshire* and *Leicestershire*, inclosing has greatly prevailed; and most of the new-inclosed lordships, says a very sensible writer, " are turned into pasturage; in consequence of " which, many lordships have not now 50 acres " ploughed yearly, in which 1500, or at least 1000 " were ploughed formerly; and scarce an ear of " corn is now to be seen in some that bore hundreds " of quarters.—And so severely are the effects of " this felt, that worse wheat has been lately sold " in these counties on an average, at 7 *s.* and " 7 *s.* 6 *d.* the *Winchester* bushel, for many months " together, than used to be sold at 3 *s.* 6 *d.* and " 4 *s.* And 5 *s.* and 5 *s.* 6 *d.* has been given for " malt that has been usually bought there at little " more than half a crown." See a pamphlet, entitled, *An Enquiry into the Reasons for and against inclosing Open Fields*, by the Rev. Mr. *Addington*. Published for Mr. *Buckland*, *Pater-noster Row*.—
In the counties of *Northampton* and *Leicester*, says the same writer, p. 43, " the decrease of the in- " habitants in almost all the inclosed villages in " which they have no considerable manufacture, " is obvious to be remarked by every one who " knew their state 20 or 30 years ago, and sees " them now; and that to a degree that cannot " but give every true friend to his country the " most sensible concern. The ruin of former " dwelling-houses, barns, stables, &c. shew every " one who passes through them that they were " once better inhabited. A hundred houses and " families have in some places, dwindled into " eight or ten.—The landholders, in most parishes " that have been inclosed only 15 or 20 years, are " very few in comparison of the numbers who " occupied them in their open field state. It is
" no

" no uncommon thing to fee four or five wealthy
" graziers engroffing a large inclofed lordfhip,
" which was before in the hands of 20 or 30 farm-
" ers, and as many fmaller tenants or proprietors.
" All thefe are hereby thrown out of their livings
" with their families, and many other families
" which were employed and fupported by them."
Ib. p. 37. See an account of *Norfolk*, in fome
refpects fimilar to this, in my *Appeal to the Public
on the Subject of the National Debt*, p. 93, &c. I
can fcarcely think of any thing that fhould be more
alarming than fuch accounts.——How aftonifhing
is it that our parliament, inftead of applying any
remedy to thefe evils, fhould chufe to promote
them, by paffing every year, bills almoft without
number, for new inclofures? (*a*)

The device, fays Lord *Bacon*, (*Effays, civil
and moral*, Sect. 20.) " of King Henry VII.
" was profound and admirable, in making
" farms and houfes of hufbandry of a ftand-

(*a*) I have here in view inclofures of *open fields and lands*
already improved. It is acknowledged by even the writers
in defence of inclofures, that thefe diminifh tillage, increafe
the monopolies of farms, raife the prices of provifions, and
produce depopulation. Such inclofures, therefore, however
gainful they may be at prefent to a few individuals, are
undoubtedly pernicious.—On the contrary, Inclofures of
wafte lands and commons would be ufeful, if divided into fmall
allotments, and given up to be occupied at moderate rents by
the poor. But if, befides leffening the produce of fine wool,
they bear hard on the poor by depriving them of a part of
their fubfiftence, and only go towards increafing farms already
too large, the advantages attending them may not much exceed the difadvantages.—He that would better inform himfelf
on this fubject, fhould, befides Mr. *Addington's* pamphlet
written *againft* inclofures, read another written *for* them, and
entitled, *The Advantages and Difadvantages of inclofing Wafte
Lands and Open Fields impartially ftated and confidered.* By a
Country Gentleman.

"ard; that is, maintained with such a pro‑
"portion of land to them, as may breed a
"subject in convenient plenty and no ser‑
"vile condition, and to keep the plough in
"the hands of the *owners* and not *hire‑
"lings*."—Inclosures, says the same great
writer, (in his History of the Reign of *Henry
the Seventh*) " began at that time (or in 1489)
"to be more frequent, whereby arable land
"was turned into pasture, which was easily
"managed by a few herdsmen. This bred
"a decay of people. In remedying this in‑
"convenience, the King's wisdom and the
"Parliament's was admirable. *Inclosures*
"they would not forbid; and *tillage* they
"would not compel; but they took a course
"to take away *depopulating inclosures*, and *de‑
"populating pasturage* by consequence. The
"ordinance was, that all houses of husban‑
"dry, with 20 acres of ground to them,
"should be kept up for ever, together with
"a competent proportion of land to be oc‑
"cupied with them, and in no wise to be
"severed from them. By these means, the
"houses being kept up, did, of necessity,
"enforce a dweller; and the proportion of
"land for occupation being also kept up,
"did, of necessity, enforce that dweller not
"to be a beggar (*a*)." The statute here
mentioned was renewed in King *Henry* the
Eighth's time; and every person who con‑

(*a*) See Lord Bacon's Works, Vol. III. p. 431.

verted tillage into pasture subjected to a forfeiture of half the land, till the offence was removed. See Mr. *Anderson's Chronological Deduction of Commerce*, Vol. I. page 347.——In a law of the 25th of the same reign, it is set forth, "that many farms, and
" great plenty of cattle, particularly sheep,
" had been gathered into few hands, where-
" by *the rents of lands had been increased,*
" *and tillage very much decayed;* churches
" and towns pulled down; the price of pro-
" visions excessively enhanced, and a mar-
" vellous number of people rendered inca-
" pable of maintaining themselves and fa-
" milies; and, therefore, it was enacted,
" that no person should keep above 2000
" sheep, *nor hold more than two farms,*"
Ib. p. 363.—In the 3d of *Edw.* VI. a bill was brought in for the benefit of the poor, for rebuilding decayed farm houses, and maintaining tillage *against too much inclosing.* Parliamentary Hist. Vol. III. p. 247.—In the year 1638, there was a special commission from *Charles* I. for enforcing the statute of the 30th of *Elizabeth*, by which no cottage was allowed in any country place, without at least four acres of land to it, to prevent the increase of the poor, by securing to them a maintenance; nor were any inmates allowed in any cottage to secure the full cultivation of the land, by diffusing the people more over it. See *Rymer's Fœd.* 20. 256. and 340.—By an

Act

Act in *Cromwell*'s time, no new house was to be built within ten miles of LONDON, unless there were four acres of land occupied by the tenant. *Parliamentary History*, Vol. XXI.

Such was the policy of former times.—*Modern* policy is, indeed, more favourable to the higher classes of people; and the consequence of it may in time prove, that the whole kingdom will consist of only *gentry* and *beggars*, or of *grandees* and *slaves*.

I cannot conclude this Postscript without adding one farther observation which has struck me on the present subject.—As in former times the numbers of the occupiers of land was greater, and all had more opportunities of working for *themselves*, it is reasonable to conclude, that the number of people willing to work for *others*, must have been smaller, and the price of day-labour higher. This is now the case in our *American* colonies; and this likewise, upon enquiry, I find to have been the case in this country formerly.——The *nominal* price of day-labour is at present no more than about *four* times, or at most *five* times higher than it was in the year 1514. But the price of corn (*a*) is *seven* times, and of flesh-meat and rayment about *fifteen* times higher. See the

(*a*) See *Chronicon Pretiosum*, Chap. V. From whence, compared with the account in Chap. IV. of the price of corn and other commodities, for the last 600 years, abundant evidence for what I have here observed, may be collected.

note,

note, p. 385.—So far, therefore, has the price of labour been from advancing in proportion to the increase in the expences of living, that it does not appear that it bears now *half* the proportion to those expences that it did formerly.

Upon the whole. The circumstances of the lower ranks of men are altered in almost every respect for the worse. From little occupiers of land, they are reduced to the state of *day-labourers* and *hirelings*; and at the same time their subsistence in that state is become more difficult, in consequence of the cause just assigned; and also of *luxury*, which has extended its influence even to them, tho' starving, and rendered *tea, fine wheaten bread*, and other delicacies, necessary to them, which were formerly unknown among them.—— Such a change cannot but draw after it important consequences. It is the lower people chiefly that pay the taxes of a state, fight its battles, carry on its commerce, and maintain its splendor. In every country, the higher ranks are a very small body, compared with them. Even in this country, where their numbers are probably much lessened, they are still more the majority than is commonly imagined; for, from the returns made by the surveyors of the house and window-duties, it appears, that THREE-FOURTHS of all the houses in the kingdom are houses not having more than *seven* windows.

Addi-

SUPPLEMENT.

Additional Observations

CONCERNING

The Schemes of the Societies for providing Annuities for Widows, and for Persons in Old Age.

THE following short and easy method has occurred to me of trying the sufficiency of all such schemes as those of the LONDON *Annuity*, and the *Laudable* Societies for the benefit of widows.

In an adequate scheme it can make no difference whether the annuities themselves are paid, or the value of them in a single payment at the time they become due.—Suppose then a society just established, consisting of 600 members, all married men at the age of 40, each of whom, besides one payment in hand, is to make an annual payment of five guineas. Suppose also their wives of the same age, and every widow to be entitled, on the day

day her husband dies, to a life-annuity of 20 *l*. the first payment to be made at the end of half a year.—Suppose further, that the society is to be kept up for ever to 600 members, by admitting new ones at the age of 40, as old ones drop off.—In the first year (according to Tables III, IV, and V. *Appendix*) twelve members, at least, will die, and leave twelve widows, each entitled to 20 *l. per annum*. The value of such an annuity to commence at the end of half a year, the age being 40, is $14\frac{1}{4}$ years purchase, by Table III. *Appendix*, reckoning interest at $3\frac{1}{2}$ *per cent*. The value, therefore, of 12 such annuities; that is, the whole amount of the sums becoming payable during the course of the first year, is 3480 *l*.—The annual contribution is 600 times 5 guineas, or 3150 *l*. and this, together with its interest for about half a year, or 3205 *l*. is all that such a society could be possessed of to bear an annual expence of 3480 *l*.—It appears, therefore, that, in order to support the expence of the supposed annuities, the annual contribution of each member ought to have been more than five guineas.

In a similar way it may be proved, that neither is such an annual contribution an adequate support to an annuity of 10 *l*. if a member lives *one* year, 15 *l*. if he lives *two* years, and 21 *l*. if he lives *three* years. This will appear from the following account; into which

which I have not taken the contributions of the firſt members at the beginning of the firſt year, becauſe I ſuppoſe them ſcarcely ſufficient to bear all the expences of management during the whole duration of the ſociety: But the firſt contributions or payments in hand, of all ſubſequent members are included, theſe being neceſſary to render the ſum of the annual contributions conſtantly 3150 *l.* as the account ſuppoſes.

3260 *l.*

3260*l*. — The Stock of the society at the end of the 2d year, being the contribution of 600 members at the end of the first year, together with the interest for a year.

Deduct 1710*l*. — The value of 12 life-annuities, of 10*l*. each, to 12 widows, aged 41, left in the course of the 2d year, at 14¼ years purchase.

Remains 1550*l*.
Add - 3260*l*. — The contribution of 600 members at the end of the 2d year, together with its interest for a year.
Add - 54*l*. — Interest at 3½ of 1550*l*. for a year.

Sum - 4864*l*. — Stock at the end of *three* years.
Deduct 2520*l*. — Value of 12 annuities, 15*l*. each, to 12 widows, aged 42, left in the course of the 3d year, at 14 years puchase.

Remains 2344*l*.
Add - 3260*l*. — Contribution, together with its interest, for the 4th year.
Add - 82*l*. — Interest of 2344*l*. for a year.

Sum - 5686*l*. — Stock at the end of *four* years.
Deduct 3465*l*. — Value of 12 annuities of 21*l*. each, to 12 widows, aged 43, left the 4th year, at 13¾ years purchase (*a.*)

Remains 2221*l*.
Add - 3260*l*. — Contribution, together with its interest, for the 5th year.
Add - 78*l*. — Interest of 2221*l*. for a year.

5559*l*. — Stock at the end of *five* years.

(*a*) A society that chose thus to pay the *values* of the annuities at the time they became due, instead of the annuities themselves, would enjoy particular advantages; for little or nothing would depend on the improvement it made of money; and time would soon determine whether it went on an adequate plan. —— A proof of the same nature with that here given,

may

SUPPLEMENT. 399

It muſt be obſerved, that the ſtock laſt given, is leſs than that immediately preceding it; and that, conſequently, in 5 years, the ſociety muſt begin to run out, and the annual contributions appear to be inſufficient.

The firſt members will leave much the ſame number of widows every year, for a

may be deduced, by conſidering theſe ſocieties as bodies of men united for the purpoſe of aſſuring to one another, from year to year, annuities for their widows; and the way of finding the value of ſuch an aſſurance is, to multiply the value of the annuity, by the probability that it will become payable in the courſe of the year.—For inſtance, Let the member's age, and alſo his wife's, be 40. Let the annuity be 20 *l. per ann.* for life, or an annuity whoſe preſent value is, by Table VI. (reckoning intereſt at $3\frac{1}{2}$ *per cent.*) 14 years purchaſe; that is, 280 *l.* The probability that a perſon at the age of 40 will die in a year, and that his wife of the ſame age will live a year; or, in other words, the probability, that ſuch a member will leave a widow in the courſe of the year, is, by Tab. III. $\frac{9}{247}$ multiplied by $\frac{436}{445}$, or .0198. (See p. 18 and 23.) That is; there will be the odds of nearly 49 to 1, againſt ſuch a member leaving a widow in the courſe of the year. The value of the aſſurance, therefore, is .0198, multiplied by 280, or the 50th part of *l.* 280; that is, 5 *l.* 11 *s.* —In the ſame manner the value of a like aſſurance for a year at any other ages may be eaſily calculated. At the age of 35, it is 5 *l.* 7 *s.* At the age of 45, it is 6 *l.* 7 *s.* The value, therefore, increaſes continually with age; and, if given in an annual payment conſtantly the ſame, which is the caſe in theſe ſocieties, it ought to be greater than the annual payment due for one year at the commencement of the aſſurance.

Five guineas *per annum*, therefore, is, *demonſtrably*, an inſufficient payment from a married man for a life-annuity of 20 *l.* to his widow.

I few

few of the first years of the scheme. After the first year also, the members admitted to supply vacancies, (about 24 annually) will begin to leave widows; and, as the whole collective body (supposed to be kept up to 600) will be continually growing older, the deaths among them, and consequently the number of widows left annually, will be continually increasing; whereas I have supposed them to remain the same.—This calculation, therefore, is as favourable as it ought to be; and every one who will examine it must be convinced, whether acquainted or not with the method of mathematically investigating the values of life-annuities depending on survivorships, that all that the societies now subsisting promise to widows more than 20*l.* or at most 20 guineas *per annum*, for an annual contribution of 5 guineas, can have no permanent support; and, if paid to *present* annuitants, must be so much taken away from some *future* annuitants. And this appears too on the suppositions, that there is no difference of age between men and their wives, that money is improved perfectly at compound interest, and that the probabilities of life among females are not higher than among males.—How melancholy then is it to think of the encouragement that has been given to these societies?—There are now in almost every part of this kingdom, some institutions or other of this kind, formed juſt as fancy has dictated, without any knowledge of the principles on which the values of life-annuities and reverſions ought to be calculated (*a*): I can, however, with pleaſure, acquaint the public, concerning the two LONDON ſocieties, of which I have taken more particular notice, that, conſiſting in general of gentlemen of character and ſenſe, they have liſtened to the information which has been offered them; and, in conſequence of it, either have already, or probably will ſoon, reſolve on ſuch amendments of their plans as may render them permanently and effectually the means of the good intended by them (*b*).

I wiſh

(*a*) There is a ſociety held at the *Nag's-head* Tavern, *Leadenhall-ſtreet*, called the AMICABLE ASSOCIATION, for the benefit of widows and children, eſtabliſhed July 7, 1767; which, for no more than an annual payment of *two guineas*, not only promiſes the very annuity mentioned above to the *widows* of members, but, if they leave no *widows*, to their *children* alſo 'till they arrive at the age of fourteen years, beſides 5 *l.* towards putting them to apprenticeſhips.——There are, I am afraid, ſeveral more ſuch *wretched* inſtitutions in LONDON; beſides many ſcattered every where in the country.

(*b*) The *London Annuity* Society, inſtead of promiſing annuities of 30 *l.* to widows, if a member lives *ſeven* years, and of 40 *l.* if he lives *fifteen* years, now offer only an advance to 30 *l. per ann.* if a member ſurvives the laſt of theſe periods. This makes a very conſiderable amendment, but it is not ſufficient; for the demonſtrations in this work, and eſpecially that in the note, page 399, may aſſure them, that their contributions will

I wish I could speak with the same satisfaction of the associations in LONDON for providing for Old Age. It is true, *they* are likewise endeavouring to reform; but in general, as far as I know any thing of them, so feebly and ineffectually as to leave little room to doubt, but they will remain what they at present undoubtedly are, SCHEMES OF FRAUD AND THEFT.——Some of them, in consequence of advancements, since the first publication of this work, require now from those who apply for admission higher contributions than those recited in the 4th Sect. Chap. II. of this work. But they ought to remember, that 'till all who have hitherto contributed too little, have either advanced their contributions and paid the compensation-money mentioned in page 116, or consented to such deductions from their annuities, as shall be proportioned to the deficiencies in their payments: They ought, I say, to remember, that 'till *this* is done, a reformation that went even so far as to require the *full* values

will bear nothing beyond the first annuity they promise, or 20*l.* if a member lives *one* year; and that as far as they give any encouragement to expect more, they raise false and unjust hopes.—The *Laudable Society* for the benefit of widows, resolved, about two months ago, at a general meeting, on a perfect reformation. But I am just now informed, that through an unhappy infatuation, they have lately revoked their resolution. I must, however, still hope, that the efforts of the wiser part of this society will some time or other meet with success.

of the annuities from all *future* members, would do them no great service.—The truth, however, is, that reckoning interest at 3½ *per cent.* their contributions are still, in general, near a *half* below what they ought to be (*a*). Is it possible then to speak of these societies with too much severity? Can any benevolent person see them, without concern, going on with schemes that have been demonstrated to be insufficient, and sure to end in confusion and calamity?—The *Provident* Society boasts, that it consists of 1280

(*a*) The true value of 30 *l. per annum*, to be enjoyed after 50, by a person now 40, is (reckoning interest at 3½ *per cent.*) 23 *l.* 10 *s.* in annual payments beginning immediately. The value required by the RATIONAL ANNUITY Society, held at the *Antwerp* Tavern, in *Threadneedle-Street*, is eight guineas in admission-money; and 4 *l.* 8 *s.* in *half-yearly* payments. This society, therefore, does not take half the value of the annuity it promises; and yet, with *singular modesty*, it assures the public, that it is formed on a plan INCONTESTABLY DURABLE.—The WESTMINSTER UNION Society of *Annuitants*, held at the *Standard* Tavern, *Leicester-Fields*, promises to a person, aged 30, an annuity of 25 *l.* for life after 48, for 3 *l.* 16 *s. per annum*, 'till 48, payable quarterly. The true value is 9 *l.* 10 *s. per annum*, payable quarterly. The value *required* by the same society at the age of 10, is 1 *l. per annum*. The *true* value is 2 *l.* 13 *s. per annum.*—Every one who will calculate in the manner directed in p. 112, &c. or in Quest. VI. p. 17. may make himself as sure of all this as he can be of any thing.

I have here mentioned the two last societies particularly, because no notice has been taken of them in page 110, &c.

members; and the *Laudable* Society, that it possesses an income of 9000 *l. per annum.*—What is this but shamelessly boasting of the numbers they have deceived, and the extensive mischief they are doing?—Some time ago they might have pleaded ignorance; but this is a plea they cannot now make.

There are FOUR societies which I must except from these censures.—The members of the FRIENDLY Society, the CONSOLIDATED Society, and the PUBLIC ANNUITANT Society, convinced of the insufficiency of their plan, have lately done themselves great honour by resolving to break up, and returning undiminished the money they had received. I have just now learnt also, that the Society of *London Annuitants*, mentioned p. 110, is come to the same resolution; and its dissolution, after some struggles, finally determined, in consequence of the zeal of many worthy and respectable members, particularly Mr. *James Palmer*, Mr. *John Chorley*, Mr. *Thomas Marsham*, Mr. *Thomas Giffin*, and the ingenious Mr. *Henley*, well known to many in the philosophical world for his skill in *Electrical* experiments.

It is necessary I should add, in order to prevent mistakes, that the society for granting *annuities increasing by survivorship* goes on a plan different from any I have considered, and the nature of which implies safety.

Some

Some think, that these societies may provide a proper security for younger members, and for all that shall become annuitants in more remote periods, by preserving untouched all the stock they shall be possessed of, at the time when the payment of the annuities shall begin. But this is a great mistake. An inadequate plan must necessarily benefit *some* by robbing *others*. For some years after the commencement of the annuities, the annual income of a society must exceed its disbursements; and all that time the first annuitants will receive more than they ought to receive, at the expence of all that are to come after them; nor is there a method possible of preventing this injustice.—The effect, in particular, of such a regulation as that now mentioned, will only be, that a *little* will be secured to annuitants in later periods, whereas otherwise they might have had *nothing*. I should be too tedious, were I to enter minutely into the explanation of this. The general reason of it is, that by paying too much to the *first* annuitants, that accumulation of stock which the calculations suppose (from *surplus* monies, while the annuitants are increasing) would be prevented; and the actual stock, in consequence of this, be rendered so much smaller than it should have been, as to leave but a small provision for the last annuitants.

In short. In such a society, the payments to annuitants would become equal to its income, long before their number rose to a *maximum*; and, therefore, if the society maintained its resolution not to enter into its stock, the annuities would, from that period, decrease continually, 'till, at last, they sunk as much *lower* than they ought to have been, as they were at first *higher*.

I have mentioned in the introduction to this work, p. 10, the interposition of the legislature. I don't know that this is to be expected. But the following reasons seem to prove that it would be proper, should any of these societies continue much longer deaf to the calls of justice and humanity.

First. They are laying (as I have proved) the foundation of much future mischief; and no government ought to see this with a careless eye.

Secondly. The principle by which they are upheld is base and iniquitous. The *present* members believe that the schemes they are supporting will last their time, and that they shall be gainers; and as for the injury done to their successors, it is at a distance, and they care little about it.—In conformity to this principle, the founders of these societies begin low; *so* low, as not to require, perhaps, a *fourth* or a *fifth* of the values of the

an-

annuities they promife. Afterwards; they advance gradually, juft as if they imagined, that the value of the annuities was nothing determinate, but increafed with every increafe of the fociety. But, as no ignorance can believe this, the true defign appears to be, to form foon as large a fociety as poffible, by leading the unwary to endeavour to be foremoft in their applications, leaft the advantage of getting in on the eafieft terms, fhould be loft.—It is well known, that thefe arts have fucceeded wonderfully, and that, in confequence of them, thefe focieties now confift of perfons who, for the *fame* annuities, make higher or lower payments according to the time when they have been admitted; and the generality of whom, therefore, muft know, that either more than the values have been required of the members laft admitted; or if not, that they are themfelves expecting confiderable annuities, for which they have given no valuable confideration, and which, if paid them, muft be ftolen from the pockets of fome of their fellow-members. What fcenes, therefore, of *difhonefty* on the one hand, and of unhappy *credulity* on the other, are thefe focieties ? (*a*).

Thirdly.

(*a*) If any perfon wants more information than I have given him concerning thefe focieties, or wifhes to fee a more ample and minute account of the infufficiency and iniquity of their fchemes, he fhould confult an ufeful work

Thirdly. There are many honest men in these societies, who having, thro' misinformation, had the misfortune to enter into them, now repent, and would be glad to withdraw. But, having made considerable payments which they cannot get back, they are forced to go on with further payments, in order to avoid losing all their former ones. These persons wish for assistance from the legislature; and their cases, I think, require assistance.

Fourthly. The sufferers by these associations may, perhaps, some time or other, come to be burdens on the public. This happened in the case of the sufferers by the CHARITABLE CORPORATION, for whose relief the parliament, in the year 1733, granted a lottery of half a million. The company of MERCERS are also now enjoying a parliamentary aid, in order to enable them to fulfil their engagements to widows; and it is well known, what expences were brought on the public by the bubbles in the *South-sea* year. —Ought not then the danger of such expences hereafter to be prevented?

work published since the last edition of this treatise, and entitled, CALCULATIONS *deduced from first Principles, in the most familiar Manner, by plain Arithmetic, for the Use of the Societies instituted for the Benefit of old Age*; intended as an *Introduction to the Study of the Doctrine of Annuities*. By a Member of one of the Societies.

After

After all. Perhaps the enforcing of an act made in the year 1720, commonly called the BUBBLE Act, might be sufficient to break these societies: And I hope that the honest part of them, should they continue to be overborn by numbers, will think, either of having recourse to this act, or of applying by petition to *Parliament*, which, when their case is in this manner brought under its cognizance, will, most probably, soon give relief.

I will add, that it seems to me, that were these societies indeed formed on durable plans, there would be reason for subjecting them to some regulations. In all of them the annuities are to commence several years before old age. Such annuities, were they to become very common in a state, might have a bad effect, by weakening the motives to industry, and promoting dissipation and idleness. —I have declared a high opinion of some institutions of this sort. Indeed no one can think more highly of them, when their object is the support of the destitute widow, or in any way the relief of unavoidable distress; and, particularly, when they are designed to enable the lower part of mankind, to provide against the wants and incapacities of old age. I have proposed a plan of this kind at the end of the third Sect. Chap. II. and I will here beg leave to recommend

another,

another, which, I think, were it carried into execution, would be very useful. I mean, a plan for establishing PARISH ANNUITIES, lately published in a pamphlet, entitled, *A Proposal for establishing Life Annuities in Parishes, for the Benefit of the industrious Poor:* Printed for Mr. *White*, in *Fleet-Street*.—" It is a common (*a*) observation," as the ingenious and public-spirited writer of this pamphlet observes, " that
" the money annually raised for the poor,
" amounts to, *at least*, a million a year;
" and that yet in many places they are
" but indifferently provided for. To make
" provision for one's old age is so na-
" tural a piece of prudence, that it seems
" at first sight wonderful, that it should not
" be generally practised by the labouring
" poor, as it is almost universally by per-
" sons in the higher paths of industry: Nor
" can their negligence in this respect be
" accounted for, in any other way so na-
" turally, as by ascribing it to their wanting
" proper opportunities of employing the
" money they might save, in some safe and

(*a*) The amount of the poor-rate for one year at the end of the reign of king Charles II. was 665,362 *l*. See *Davenant*'s works, Vol. I. p. 38.——The prices of the means of subsistence have been since doubled; and when this is considered; and also, that an increase of parish poor must be one effect of the causes that produce depopulation; it will appear probable, that the observation above-mentioned does not exceed the truth.

" easy

" eafy method that would procure them a
" fuitable advantage from it in the latter pe-
" riods of their lives. They know, for the
" moſt part, but little of the *public funds*;
" and when it happens that they are ac-
" quainted with them, the fmallnefs of the
" fums they would be entitled to receive, as
" the intereſt of the money they could af-
" ford to lay out in them, is no encourage-
" ment to them to difpofe of it in that way.
" What inducement, for inſtance, can it be
" to a man who has faved ten pounds out
" of his year's wages, to inveſt it in the pur-
" chafe of 3 *per cent. Bank annuities*, to con-
" fider that it will produce him fix or feven
" fhillings a year? It is but the wages of
" three days labour.—And if they lend their
" money to tradefmen of their acquaintance,
" as they fometimes do, it happens not un-
" frequently that their creditor becomes a
" bankrupt, and the money they had truſted
" him with is loſt for ever; which difcou-
" rages others of them from faving their
" money at all, and makes them refolve to
" fpend it in the enjoyment of prefent plea-
" fure. But if they faw an eafy method of
" employing the money they could fpare, in
" fuch a manner as would procure them a
" confiderable income in return for it at fome
" future period of their lives, without any
" fuch hazard of lofing it by another man's
                                    " folly

" folly or misfortune, it is probable they
" would frequently embrace it: And thus a
" diminution of the poor rate on the estates
" of the rich, an increase of present industry
" and sobriety in the poor, and a more in-
" dependent and comfortable support of them
" in their old age, would be the happy con-
" sequences of such an establishment. Now
" this might be effected in the following
" method.

First, " Let the church-wardens and over-
" seers of every parish be impowered, by act
" of parliament, to grant life-annuities to
" such of the inhabitants of the parish, as
" shall be inclined to purchase them, to com-
" mence at the end of one, two, or three
" years, or such other future period of time
" as the purchaser shall chuse, and to be paid
" out of the poor rates of the parish, so that
" the lands and other property in the parish
" that is chargeable to the poor-rate, shall
" be answerable for the payment of these
" annuities.—This circumstance would give
" these annuities great credit with the poor
" inhabitants, by setting before them a so-
" lid and ample security for the payment of
" them.

Secondly, " Let the annuities, thus grant-
" ed to the poor inhabitants, be such as arise
" from a supposition that the interest of mo-
" ney is 3 *per cent.* or some higher rate of
" interest,

" intereſt, if the churchwardens and over-
" ſeers of the poor think fit to make uſe of
" ſuch higher intereſt.

Thirdly, " But at the rate of 3 *per cent.*
" the purchaſer ſhould have a right to an an-
" nuity, and the church-wardens and over-
" ſeers of the poor ſhould be compellable to
" grant it.

Fourthly, " No annuity depending on one
" life ſhould exceed 20 *l.* a year.

Fifthly, " No leſs ſum than 5 *l.* ſhould
" be allowed to be employed in the pur-
" chaſe of an annuity.——This is to avoid
" intricacy and multiplicity in the accounts.

Sixthly, " An exact regiſter of theſe grants
" ſhould be kept, by the church-wardens and
" overſeers of the poor, in proper books for
" the purpoſe, in which the grants ſhould
" be copied exactly, and the copy of each
" grant ſubſcribed by the perſon to whom it
" is granted. And this copy, in the regiſter-
" book of the pariſh, ſhould be good evi-
" dence of the purchaſer's right to the an-
" nuity, in caſe the original deed of grant to
" the purchaſer, which was delivered to
" him at the time of the purchaſe, ſhould
" be afterwards loſt.

Seventhly, " The money thus paid to the
" church-wardens and overſeers of the poor
" for the purpoſe of life-annuities, ſhould
" be employed in the purchaſe of 3 *per cent.*
" Bank-

"Bank-annuities in the joint names of all the church-wardens and overseers, and by them transferred at the expiration of their offices to their successors, and so on to the next successors for ever, so as to be always the legal property of the church-wardens and overseers of the poor for the time being, in trust for the persons who should be entitled to the several life-annuities, granted in the manner above-mentioned; and the interest of this money should be received every half year, and invested in the purchase of more principal continually, so as to make a perpetual fund for the payment of the annuities, &c. &c. Deficiencies, if any should ever happen, to be made good by the poor-rates, &c. &c."

I hope I shall be excused the length of this Quotation. The particulars recited in it are followed, by an account of the annuities to which the payment of 10 *l.* at the age of 25, would entitle, after attaining to the age of 30, 35, 40, 45, &c. and also by a very just and clear explanation of the method of calculating such annuities.

To the whole is added, a draught of an Act of Parliament for enabling parish-officers to grant such annuities, drawn up in consequence of instructions from some members of the House of Commons, and particularly

one

one gentleman of great eminence, who has signified an intention of bringing such a bill into parliament.

I have no alterations in this scheme to propose, that I think very material. I rejoice to find that it is likely to be brought under the consideration of the legislature. I will, however, just mention, that in order to avoid all danger of checking industry among the poor, it would, perhaps, be right to provide that the annuities shall not commence before the purchaser has compleated the age of 50, 55 or 60? And also, that it might be best, that the annuities should be made to increase gradually with the increasing infirmities of age, till they became greatest at 65 or 70 years of age, when their aid will be most wanted?

For instance. Let the annuity begin with 10 *l.* for 5 years. At the end of 5 years, let it rise to 20 *l.* for five years more; and after that let it be 30 *l.* for the whole remainder of life. Let also every purchaser be allowed to chuse at what age his annuity shall commence; and, as a further advantage, let it be payable *quarterly*, and let him be allowed to purchase $\frac{1}{2}$, $\frac{1}{3}$, $\frac{1}{4}$, &c. of the annuity, just as he shall like or can best afford.—In this way, persons who are now young might make an ample provision for old age on very easy and inviting terms.

SUPPLEMENT.

A respectable body of men in this kingdom, whose subsistence too generally depends on the continuance of their capacities of present service; have, for some time, had under consideration a plan of this sort; and a set of tables has been composed for them. As possibly these tables may be of some general use, I shall beg leave to subjoin them.

TABLE I.

Shewing the present Value of an Annuity of 10*l.* for five years; 20*l.* for the next succeeding five years; and 30*l.* for the whole of life after ten years; payable *quarterly*; and to commence at FIFTY-FIVE years of age.

Age of the Purchaser.	Value of the Annuity in one present Payment.		Value of the annuity in annual payments, to be continued 'till 55, the 1st payment to be made immediately.	
	l.	*s.*	*l.*	*s.*
20	38	6	2	4
21	40	7	2	7
22	42	8	2	10
23	44	9	2	13
24	46	11	2	16
25	48	13	3	0
26	51	3	3	4
27	53	14	3	8
28	56	6	3	13
29	58	18	3	18
30	61	11	4	4
31	64	16	4	11
32	68	1	4	18
33	71	7	5	5
34	74	13	5	13
35	78	0	6	1
36	81	16	6	11
37	85	12	7	2
38	89	9	7	13
39	94	0	8	6
40	98	11	9	0
41	103	16	10	0
42	109	0	11	0
43	114	4	12	3
44	121	0	13	13
45	128	8	15	9

SUPPLEMENT.

TABLE II.

Shewing the Values of an Annuity of 10 *l.* for five years; 20 *l.* for the next succeeding five years; and 30 *l.* for the whole of life after ten years; payable *quarterly*, and to commence at SIXTY years of Age.

Age of the Purchaser.	Value of the Annuity in one present Payment.		Value of the Annuity in annual Payments, to be continued till the Age of 60, the first Payment to be made immediately.	
	l.	*s.*	*l.*	*s.*
20	22	13	1	5
21	23	18	1	6
22	25	3	1	8
23	26	8	1	10
24	27	13	1	12
25	28	19	1	14
26	30	10	1	16
27	32	2	1	18
28	33	13	2	0
29	35	4	2	3
30	36	18	2	6
31	38	12	2	9
32	40	8	2	12
33	42	5	2	15
34	44	2	2	19
35	46	0	3	3
36	48	10	3	8
37	51	0	3	13
38	53	10	3	19
39	56	5	4	5
40	59	0	4	12
41	61	10	5	0
42	64	10	5	8
43	68	0	5	18
44	72	10	6	14
45	77	0	7	10
46	81	10	8	4
47	86	0	9	0
48	90	10	9	16
49	96	0	11	0
50	102	0	12	10

SUPPLEMENT. 419
TABLE III.

Shewing the Values of an Annuity of 10 *l.* for five years; 20 *l.* for the next succeeding five years; and 30 *l.* for the whole of life after *ten* years; payable *quarterly*, and to commence at SIXTY-FIVE years of age.

Age of the Purchaser.	Value of the Annuity in one present Payment.		Value of the Annuity in annual Payments, to be continued till the Age 65, and to begin immediately.		
	l.	*s.*	*l.*	*s.*	*d.*
20	12	4	0	13	0
21	12	18	0	13	9
22	13	12	0	14	6
23	14	6	0	15	6
24	15	1	0	16	6
25	15	16	0	17	6
26	16	13	0	18	6
27	17	10	0	19	6
28	18	7	1	0	6
29	19	5	1	2	0
30	20	3	1	3	6
31	21	5	1	5	0
32	22	7	1	7	0
33	23	10	1	9	0
34	24	13	1	11	0
35	25	16	1	13	0
36	27	0	1	15	0
37	28	4	1	17	6
38	29	9	2	0	0
39	30	14	2	3	0
40	32	0	2	6	0
41	34	0	2	10	0
42	36	0	2	14	0
43	38	0	2	18	0
44	40	0	3	3	0
45	42	5	3	8	0
46	44	15	3	14	0
47	47	9	4	1	0
48	50	3	4	9	0
49	53	0	4	18	0
50	55	18	5	7	0
51	59	0	6	0	0
52	62	10	6	15	0
53	67	0	7	10	0
54	72	0	8	10	0
55	77	12	9	17	0

SUPPLEMENT.

These TABLES have been calculated by the rules in Queſt. VI. page 17, 18, &c. The probabilities of life have been taken from Table IV. page 323: And the intereſt of money reckoned at 3 *per cent*.

It is proper, in order to prevent all danger of miſtakes, to add, that the values in each of the ſecond and third columns of theſe Tables, are the *whole* values. That is, The values in the *ſecond* column of every Table ſuppoſe the payments in the third column excuſed. And, in like manner, the values in the *third* column ſuppoſe the payments in the ſecond excuſed.

TABLE

SUPPLEMENT. 421
TABLE IV. (a)

Shewing the Probabilities of Life in the District of VAUD, SWITZERLAND, formed from the Registers of 43 Parishes, given by Mr. *Muret*, in the First Part of the BERN Memoirs for the Year 1766.

Age.	Living	Decr.	Age.	Living	Decr.	Age.	Living	Decr.
0	1000	189	31	558	5	62	286	12
1	811	46	32	553	5	63	274	12
2	765	30	33	548	4	64	262	12
3	735	20	34	544	5			
4	715	14				65	250	14
			35	539	6	66	236	16
5	701	13	36	533	6	67	220	18
6	688	11	37	527	7	68	202	18
7	677	10	38	520	7	69	184	16
8	667	8	39	513	7			
9	659	6				70	168	15
			40	506	6	71	153	13
10	653	5	41	500	6	72	140	11
11	648	5	42	494	6	73	129	10
12	643	4	43	488	6	74	119	10
13	639	4	44	482	6			
14	635	4				75	109	11
			45	476	7	76	98	13
15	631	5	46	469	8	77	85	14
16	626	4	47	461	10	78	71	13
17	622	4	48	451	10	79	58	12
18	618	4	49	441	10			
19	614	4				80	46	10
			50	431	9	81	36	7
20	610	4	51	422	8	82	29	5
21	606	4	52	414	8	83	24	4
22	602	5	53	406	9	84	20	3
23	597	5	54	397	9			
24	592	5				85	17	3
			55	388	11	86	14	3
25	587	5	56	377	13	87	11	2
26	582	5	57	364	16	88	9	2
27	577	5	58	348	17	89	7	2
28	572	5	59	331	17			
29	567	4				90	5	1
			60	314	15			
30	563	5	61	299	13			

(*a*) All the Bills, from which this and the following Tables are formed, give the numbers dying under 1 as well as under 2 years; and, in the numbers dying under 1, are included, in the country parish in *Brandenburg*, and at *Berlin*, all the still-borns. All the bills also give the numbers dying in every period of five years.

TABLE V.

Shewing the Probabilities of Life in a Country Parish in BRANDENBURG, formed from the Bills for 50 Years, from 1710 to 1759, as given by Mr. SUSMILCH, in his *Gottliche Orduung*, page 43.

Age.	Living.	Decr.	Age.	Living.	Decr.	Age.	Living.	Decr.
0	1000	225	31	482	5	62	260	12
1	775	57	32	477	5	63	248	12
2	718	31	33	472	5	64	236	12
3	687	23	34	467	5	65	224	11
4	664	22	35	462	6	66	213	11
5	642	20	36	456	6	67	202	12
6	622	15	37	450	6	68	190	12
7	607	12	38	444	6	69	178	12
8	595	10	39	438	6	70	166	13
9	585	8	40	432	5	71	153	15
10	577	7	41	427	5	72	138	16
11	570	6	42	422	5	73	122	15
12	564	5	43	417	5	74	107	14
13	559	5	44	412	6	75	93	13
14	554	5	45	407	6	76	80	12
15	549	5	46	400	6	77	68	9
16	544	5	47	394	6	78	59	8
17	539	4	48	388	7	79	51	7
18	535	4	49	381	7	80	44	6
19	531	4	50	374	7	81	38	6
20	527	5	51	367	8	82	32	6
21	522	5	52	359	8	83	25	6
22	517	5	53	351	8	84	21	5
23	512	5	54	343	9	85	15	4
24	507	5	55	334	10	86	11	3
25	502	4	56	324	10	87	8	2
26	498	3	57	314	10	88	6	2
27	495	3	58	304	11	89	4	1
28	492	3	59	293	11	90	3	1
29	489	3	60	282	11	91	2	1
30	486	4	61	271	11	92	1	1

SUPPLEMENT. 423
TABLE VI.

Shewing the Probabilities of Life in the Parish of HOLY-CROSS, near SHREWSBURY, formed from a Register kept by the Rev. Mr. *Gorsuch*, for 20 years, from 1750 to 1770. See Page 192, 259, 263.

Age.	Living.	Decr.	Age.	Living.	Decr.	Age.	Living.	Decr.
0	1000	178	31	481	5	62	253	10
1	882	60	32	476	5	63	243	10
2	762	45	33	471	5	64	233	10
3	717	35	34	466	6	65	223	10
4	682	23	35	460	6	66	213	10
5	659	23	36	454	7	67	203	10
6	636	18	37	447	7	68	193	11
7	618	14	38	440	7	69	182	11
8	604	9	39	433	7	70	171	10
9	595	6	40	426	8	71	161	10
10	599	4	41	418	8	72	151	9
11	585	4	42	410	9	73	142	8
12	581	4	43	401	8	74	134	8
13	577	4	44	393	7	75	126	7
14	573	4	45	386	7	76	119	7
15	569	4	46	379	7	77	112	7
16	565	5	47	372	7	78	105	7
17	560	5	48	365	6	79	98	8
18	555	5	49	359	6	80	90	9
19	550	5	50	353	6	81	81	10
20	545	6	51	347	7	82	71	10
21	539	7	52	340	7	83	61	10
22	532	7	53	333	7	84	51	10
23	525	7	54	326	8	85	41	9
24	518	6	55	318	8	86	32	8
25	512	6	56	310	9	87	24	7
26	506	5	57	301	9	88	17	6
27	501	5	58	292	9	89	11	4
28	496	5	59	283	10	90	7	2
29	491	5	60	273	10	91	5	1
30	486	5	61	263	10	92	4	1

TABLE VII.

Shewing the Probabilities of Life at VIENNA, formed from the Bills for Eight Years, as given by Mr. SUSMILCH, in his *Gottliche Ordnung*, Page 32, Tables.

Age.	Living.	Decr.	Age.	Living.	Decr.	Age.	Living.	Decr.
0	1495	682	31	364	6	62	129	6
1	813	107	32	358	5	63	123	7
2	706	61	33	353	6	64	116	7
3	645	46	34	347	7	65	109	8
4	599	33	35	340	8	66	101	8
5	566	30	36	332	8	67	93	8
6	536	20	37	324	8	68	85	7
7	516	11	38	316	9	69	78	7
8	505	9	39	307	9	70	71	6
9	496	7	40	298	8	71	65	5
10	489	6	41	290	7	72	60	5
11	483	5	42	283	6	73	55	4
12	478	5	43	277	6	74	51	4
13	473	6	44	271	7	75	47	5
14	467	6	45	264	8	76	42	5
15	461	6	46	256	9	77	37	5
16	455	7	47	247	9	78	32	5
17	448	6	48	238	9	79	27	4
18	442	6	49	229	9	80	23	3
19	436	6	50	220	8	81	20	2
20	430	5	51	212	7	82	19	2
21	425	5	52	205	7	83	16	2
22	420	5	53	198	7	84	14	2
23	415	6	54	191	7	85	12	2
24	409	6	55	184	8	86	10	2
25	403	6	56	176	8	87	8	2
26	397	6	57	168	9	88	6	2
27	391	7	58	159	8	89	4	1
28	384	7	59	151	8	90	3	1
29	377	7	60	143	7	91	2	1
30	370	6	61	136	7	92	1	1

SUPPLEMENT. 425
TABLE VIII.

Shewing the Probabilities of Life at BERLIN, formed from the Bills for Four Years, from 1752 to 1755, given by Mr. SUSMILCH (a), in his *Gottliche Ordnung*, Vol. II. page 37, Tables.

Age.	Living	Decrs.	Age.	Living	Decrs.	Age.	Living	Decr.
0	1427	524	33	361	7	65	112	6
1	903	151	34	354	7	66	106	7
2	752	61				67	99	7
3	691	73	35	347	8	68	92	6
4	618	45	36	339	9	69	86	6
			37	330	10			
5	573	21	38	320	10	70	80	6
6	552	15	39	310	10	71	74	6
7	536	13				72	68	6
8	523	9	40	300	10	73	62	5
9	514	7	41	290	9	74	57	5
			42	281	8			
10	507	5	43	274	7	75	52	5
11	502	4	44	266	7	76	47	5
12	498	4				77	42	5
13	494	4	45	259	7	78	37	5
14	490	4	46	252	7	79	32	4
			47	245	7			
15	486	4	48	238	7	80	28	4
16	482	5	49	231	7	81	24	3
17	477	5				82	21	2
18	472	5	50	224	7	83	19	2
19	467	6	51	217	7	84	17	2
			52	210	7			
20	461	6	53	203	8	85	15	2
21	455	6	54	195	8	86	13	2
22	449	6				87	11	2
23	443	7	55	187	8	88	9	2
24	436	8	56	179	8	89	7	1
			57	171	8			
25	428	9	58	163	9	90	6	1
26	421	9	59	154	9	91	5	1
27	412	9				92	4	1
28	403	9	60	145	8	93	3	1
29	394	9	61	137	7	94	2	1
			62	130	6			
30	385	9	63	124	6			
31	376	8	64	118	6			
32	368	7						

(a) This writer has also given the bills of the parish of St. *Peter's* at BERLIN, for 24 years; and a Table formed from them, agrees nearly with this.

The

The following facts came to my knowledge too late to be inserted in their proper places. They furnish additional evidence for some of the observations I have made; and, therefore, I have chosen to introduce an account of them here, rather than entirely omit them.

An exact account was taken in August, 1772, by the desire of the Earl of *Shelburne*, of the number of families, and of inhabitants in CALNE, a manufacturing town in Wiltshire.—The number of *married* persons and heads of families was 1102; of *single* heads of families, 241; of children, 1614; of lodgers and servants, 510; of families, 776; and of inhabitants of all ages and conditions, exclusive of 58 in the poor-house, 3467; or near 4¼ to a *family*.

About the same time an exact account was taken also of the town and parish of WYCOMBE in *Buckinghamshire*, and the number of families in the *town* was found to be 432; and of inhabitants, exclusive of 46 in the poor-house, 2152, or 5 to a *fa-*

The numbers *born* at BERLIN, during the 4 years abovementioned, were, *males*, 9219; *females*, 8743; or 21 to 20.

The numbers that died under 2 years of age, were, *males*, 3118; *females*, 2623; or 7 to 6.

The numbers that died upwards of 80 years of age, were, *males*, 135; *females*, 215; or 5 to 8.

The numbers that died between 91 and 105, were, *males*, 21; *females*, 55.

mily.

mily. In that part of the parish which lies in the country, were 68 families, and 309 inhabitants, or $4\frac{1}{2}$ to a *family*.

At ALTRINGHAM, a market-town in Cheshire, according to an accurate survey made in July last, the number of *houses* was 248, of inhabitants, 1029; or $4\frac{1}{7}$ to a *house*.

St. *Michael's*, a small parish in the center of the town of CHESTER, contains, according to a very exact account taken under the direction of Dr. HAYGARTH, 246 males, 372 females, 166 married persons, 41 widows, 21 widowers, 137 children under 15 years of age, 151 families, 127 houses, and 618 inhabitants, or $4\frac{1}{11}$ to a *family*, and $4\frac{1}{4}$ to a *house*.

At BIRMINGHAM, in the year 1700,
The *inhabitants* were 15032
The *houses* — — 2504, or 6 to a *house*.

In 1750,
The *inhabitants* were 23688
The *houses* — — 4170, or $5\frac{7}{10}$ to a *house*.

In 1770,
The *males* were — 15363
The *females* — — 15441

Total of Inhabitants in 1770—30804
    Houses — — — 6025, or $5\frac{1}{4}$ to a house.

We may fee, in this account, the progrefs of luxury at BIRMINGHAM; the houfes there having increafed fo much fafter than the inhabitants, that 600 houfes now contain no more people than 511 contained 70 years ago.

In a *hundred* fmall towns and parifhes in the generality of ROUEN, 26 in the generality of LYONS, and 16 in the generality of AUVERGNE in FRANCE, the *married* men and widowers were a few years ago 19916; the married women and widows 22494; the males 47817; the females 51185; the inhabitants of all ages and conditions 99002; the families, 24910, or nearly 4 to a family. See *Recherches fur la Population,* par M. Meffance, page 8, 26, 62.

Similar accounts of *Norwich, Manchefter, Leeds, Shrewfbury, Northampton, Newbury, Rome,* the diftrict of *Vaud* in *Switzerland,* &c. &c. may be found in page 183, &c. and the beginning of the Supplement.

At GAINSBROUGH, in *Lincolnfhire,* a regifter has been kept for many years of the *chriftenings, weddings,* and *burials,* in which are particularly diftinguifhed the numbers of each fex dying at every age in every month. I have lately obtained, through the affiftance of a friend who lives in this town, a copy of this regifter for 20 years back, or from 1752 to 1771.——The annual medium of chriften-

chriftenings during this period, including all among diffenters, has been 126; of weddings, 34; of burials, 105.—The *weddings* in fummer (July, Auguft, September) have been 130. In winter (December, January, March) 144. In autumn, 188. In fpring, 218.——The *chriftenings* in fummer (June, July, Auguft, and September) have been 779. In winter (December, January, February, March) 811.——The *burials* in the fame four fummer months, have been 590. In the four winter months, 765. The mortality of *fummer*, therefore, in this town, is lefs than the mortality of *winter*, in the proportion of 40 to 52. See the note in p. 371. The burials in *April* and *May* have been 390. In *October* and *November*, 345.—The chriftenings in *April* and *May* have been 427. In *October* and *November*, 410

Died

At GAINSBROUGH,	Males.	Females.	Both sexes.
Died under 20	525	485	1010
Between 20 and 25	32	39	71
25 and 30	25	41	66
30 and 35	30	41	71
35 and 40	28	35	63
40 and 45	35	30	65
45 and 50	35	25	60
50 and 55	47	48	95
55 and 60	53	49	102
60 and 65	57	73	130
65 and 70	43	50	93
70 and 75	51	51	102
75 and 80	31	30	61
80 and 101	32	49	81
Of all ages in 20 years	1024	1046	2070

According to this Table, one-half of all that are christened live to 22 years of age; and 81 of 2070, that is 1 in 25¼, live to 80, of whom the major part, in the proportion of 49 to 32, are females.

The town and parish of GAINSBROUGH consist of 920 houses; of which 161 are houses in the hamlets and country round the town.

A TABLE

SUPPLEMENT.

A TABLE shewing the numbers who have died at all ages for 10 years, in two towns, and 13 parishes, in the generalities of *Lyon* and *Rouen* in *France*. Taken from *Recherches sur la Population*, &c. par M. *Messance*.

```
Died under  5 ——— 2167
From  5 to 10  ——  290
     10 to 20  ——  279
     20 to 30  ——  309
     30 to 40  ——  307
     40 to 50  ——  297
     50 to 60  ——  315
     60 to 70  ——  341
     70 to 80  ——  364
     80 to 90  ——  195
     90 to 100 ——   22
                  ————
                  4884
```

FINIS.

www.ingramcontent.com/pod-product-compliance
Lightning Source LLC
Chambersburg PA
CBHW030322020526
44117CB00030B/566